Experimental
Developmental
Biology

Experimental Developmental Biology

A Laboratory Manual

Laura R. Keller

Department of Biological Science
Florida State University
Tallahassee, Florida

John H. Evans

Department of Physiology
University of Massachusetts Medical Center
Worcester, Massachusetts

Thomas C. S. Keller

Department of Biological Science
Florida State University
Tallahassee, Florida

Academic Press

San Diego   London   Boston   New York   Sydney   Tokyo   Toronto

Front cover photograph: (Background) Newly fertilized eggs and
2- , 4- , and 8-cell sea urchin embryos of the species *Lytichinus pictus*.
(Inset) A *Drosphila* larva (bottom), pupa (center), and adult fly (top).
Both micrographs taken by Laura R. Keller.

This book is printed on acid-free paper. ∞

Copyright © 1999 by ACADEMIC PRESS

Academic Press
a division of Harcourt Brace & Company
525 B Street, Suite 1900, San Diego, California 92101-4495, USA
http://www.apnet.com

Academic Press
24-28 Oval Road, London NW1 7DX, UK
http://www.hbuk.co.uk/ap/

International Standard Book Number: 0-12-403970-7

PRINTED IN THE UNITED STATES OF AMERICA
98 99 00 01 02 03 EB 9 8 7 6 5 4 3 2 1

CONTENTS

PREFACE   ix

1

Sea Urchin Fertilization

I. Introduction   1
II. Experimental Protocol   4
III. Questions   9
IV. Supplementary Readings   11

2

Early Amphibian Development

I. Introduction   13
II. Experimental Protocol   14
III. Questions   19
IV. Supplementary Readings   20

3

Embryonic Chick Development

I. Introduction   21

II. Experimental Protocol   22
III. Questions   30
IV. Supplementary Readings   31

4

Drosophila Gene Expression

I. Introduction   33
II. Experimental Protocol   34
III. Questions   40
IV. Supplementary Readings   40

5

Amphibian Metamorphosis

I. Introduction   41
II. Experimental Protocol   42
III. Questions   45
IV. Supplementary Readings   45

6

Cell–Cell Interactions during Sponge Aggregation

I. Introduction   47
II. Experimental Protocol   48
III. Questions   51
IV. Supplementary Readings   52

APPENDIX A

Laboratory Requirements   53

Materials Needed by Each Student   53
Keeping a Lab Notebook   53
Writing Lab Reports   54

APPENDIX B

Care and Use of Microscopes   55

Care and Use of Microscopes   55
Workshop 1: Determination of Field Diameters   56
Workshop 2: Size Estimation   58
Workshop 3: Determination of Volumes   59

APPENDIX C

Gel Electrophoresis   61

Introduction   61
Exercises   62
Workshop 1: Plasmid Mapping   70

APPENDIX D

Guide to Preparation for Laboratory Exercises   73

   I. General List of Equipment and Supplies for
      Laboratory Exercises   73
  II. Guide to Preparation, Laboratory Exercise 1: Sea Urchin
      Fertilization   76
 III. Guide to Preparation, Laboratory Exercise 2: Early
      Amphibian Development   84
  IV. Guide to Preparation, Laboratory Exercise 3: Embryonic Chick
      Development   89
   V. Guide to Preparation, Laboratory Exercise 4: *Drosophila* Gene
      Expression   95
  VI. Guide to Preparation, Laboratory Exercise 5: Amphibian
      Metamorphosis   99
 VII. Guide to Preparation, Laboratory Exercise 6: Cell–Cell Interactions
      during Sponge Aggregation   101
VIII. Guide to Preparation, Appendix B: Microscope Care
      and Use   104
  IX. Guide to Preparation, Appendix C: Gel Electrophoresis   106

INDEX   113

PREFACE

An understanding of classical embryology is crucial for understanding modern developmental biology, but let's face it: poring over sections and whole mounts of even one's favorite animal makes it easy to miss one of the most important aspects of developmental biology—the dynamics of change in a developing organism. From formation of an embryo by fusion of a sperm and egg to completion of the adult body plan, individual cells and tissues are constantly moving, changing shape, and stimulating other cells and tissues to embark on new activities according to some specific but seemingly mysterious plan. Full appreciation of developmental processes requires observation and experimentation with live cells and organisms.

The experimental exercises in this manual are designed to give the student greater appreciation of some of these dynamic aspects of development. The developing animal systems used here are discussed in all major developmental biology texts, and their study will complement and reinforce the major concepts of a lecture course in developmental biology. Each exercise illustrates several major concepts. The first focuses on fertilization in sea urchins, in which students examine gamete morphology and probe species specificity, ionic requirements, changes in gene expression with development, and blocks to polyspermy. In studying amphibian development in the second, students examine early cleavage, totipotency of early blastomeres, and establishment of the embryonic axes. The third exercise focuses on tissue and organ formation through examination of limb bud grafting and cardia bifida (split heart) formation in avian embryos. Fourth, using *Drosophila,* students will examine morphological changes in polytene-chromosome puffing patterns and, using gel electrophoresis, molecular changes in gene expression. Cell–cell communication is illustrated in the fifth exercise, on amphibian metamorphosis, and in the sixth, which focuses on sponge-cell reaggregation.

These exercises are designed for use in 3-hour laboratory periods during which 20 students usually work in pairs. The sea urchin, frog, chick, and *Drosophila* exercises require more than one scheduled lab meeting. Because embryos continue developing after the 3 hours have past, all exercises also require additional lab visits for observation or minor work on days following the lab period. Materials covered in Appendixes A and B are suitable for use as the introductory/organizational lab. Although the exercises are presented in a "developmentally correct" sequence, the order can be changed to accommodate organism availability or to coordinate the lab exercises with lecture topics.

Included in this manual is an extensive set of instructions for lab preparation. A suggested timeline for preparation for each exercise is included, as are sources from which materials and organisms can be obtained. We have made these instructions as complete as possible, but experience and planning by the instructor will help to ensure success of the exercises.

Students observing normal development of the embryos used in these exercises will be exposed to a whole new fascinating world. As in all experimental work involving live specimens, however, success is never guaranteed. Although this uncertainty may frustrate some if not most students, it is important for them to "live," if only for 3 hours a week, the life of a research scientist. From this experience, students will gain a better appreciation of the work that has led to the "facts" in textbooks and will better understand the nature of science and the origin of those facts. Some students will rise to the challenge presented by these exercises and be thrilled when they succeed in producing a bifurcated heart, a two-headed embryo, or beautifully spread polytene chromosomes.

The authors thank all of the students who have enjoyed the frequent successes and suffered through the disasters over the 10 years that these labs have been refined. Special appreciation is due to Drs. A. Jacobson, G. Freeman, and C. Emerson for introducing one of us (L.R.K.) to their areas of developmental biology. We also thank J. Brooks, Dr. A. Johnson, and Dr. B. Mitchell for help with some of the labs, Dr. A. Thistle for superb editing work, and the editors at Academic Press for their patience and encouragement. We sincerely hope that instructors of developmental biology exercises will find this manual useful in imparting to students the beauty and excitement of working with live, developing embryos.

1 LABORATORY EXERCISE

SEA URCHIN FERTILIZATION

I. INTRODUCTION

Studies of sea urchin gametes led to the first observations of fertilization in the late 1800s. Since then, experiments with sea urchins have progressed from the classical microscopic observations of fertilization and development to more extensive studies on all aspects of their cell biology using modern biochemical and molecular techniques. Today, sea urchins remain important and productive subjects of investigations worldwide.

The popularity of sea urchins in fertilization studies is due in part to their reproductive attributes. Fertilization and development occur externally, in open seawater, therefore allowing for straightforward laboratory duplication of conditions. Female urchins produce hundreds of millions of eggs in a season; males produce hundreds of billions of sperm. Urchin eggs are easily collected by a variety of means and, if desired, can be fertilized simply by mixing with sperm in seawater. The large size of the eggs and of the resulting zygotes allows for detailed observations of their development with a microscope. After fertilization, the resulting sea urchin culture develops synchronously through early cleavage to the larval stage, making possible studies at all stages of early development (Table 1.1).

TABLE 1.1

Timetable for Normal Embryonic Development of Sea Urchins

Stage	Time after fertilization at 20°C
1-cell, unfertilized	0 min
1-cell, fertilization envelope elevated	2 min
2-cell	70–90 min
4-cell	90–110 min
Blastula	6 h
Hatched blastula	10 h
Gastrula	12–20 h
Early pluteus larva	36 h

Experimental Rationale

Sea urchin eggs are very large, and in some species translucent, and are surrounded by a transparent jelly coat. In Part II.A, you will determine the general physical characteristics (morphology, size, and volume) of the eggs and the thickness of their jelly coat.

Urchin sperm cells differ greatly from eggs in their morphology. They also exhibit a chemotactic response to certain factors in their environment. You will study these morphologic differences and the chemotactic response of the sperm in Part II.B of this exercise.

A dramatic visible event following soon after sperm penetration of the egg is the release and elevation of the vitelline membrane from the egg surface to form the fertilization membrane (Fig. 1.1). A later event following fertilization is cytokinesis. Various cations normally found in seawater are required for certain cellular events in sea urchin gametes and fertilized eggs. In Part II.C, you will treat eggs and sperm with normal seawater (SW), seawater without sodium ions (sodium-free SW or NaFSW), and seawater without calcium or magnesium ions (CaMgFSW) and observe the effects of these solutions on gametes and fertilization. From the results, you will be able to determine the roles of the ions in normal fertilization.

Polyspermy is fertilization of a single egg by more than one sperm. It leads to genetic anomalies, and eggs normally have mechanisms to prevent its occurrence. In Part II.D, you will determine the nature of one of these mechanisms by experimentally inducing polyspermy and observing the consequences of polyspermy in development.

Because sea urchin eggs are fertilized externally in open seawater, a sperm from one species may find its way to the egg of another. Evolution has provided means to prevent fertilization from occurring in this situation. The experiment in Part II.E will give you insight into the nature of cross-species fertilization prevention.

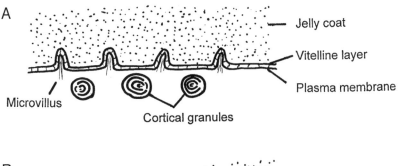

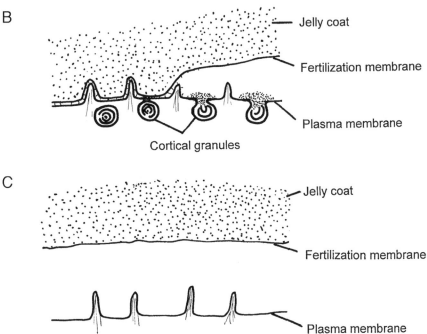

FIGURE 1.1   Sea urchin egg and embryo surfaces. A. The plasma membrane of the unfertilized egg is covered by the vitelline layer and jelly coat. B. After fertilization, connections between the plasma membrane and vitelline layer are broken, and the released vitelline layer rises above the plasma membrane, forming a physical barrier to sperm penetration, which is called the fertilization membrane.

Parthenogenesis is the development of an organism from an unfertilized egg. It can be experimentally induced by various means. By inducing parthenogenetic activation in Part II.F, you will be able to learn what cellular events accompany normal fertilization.

Sea urchin gametes and embryonic cells express a variety of proteins at different times during their development. These proteins may differ between species or be common to many species. In Part II.G of this exercise, you will determine the protein complement of the gametes and embryos of two sea urchin species by SDS–PAGE and compare proteins present in the gametes and different developmental stages of the two species.

II. EXPERIMENTAL PROTOCOL

CAUTION: Please tell the instructor if you are allergic to penicillin or streptomycin. The unpolymerized acrylamide used in Part II. G is a neurotoxin. Do not attach or detach the leads from the gel apparatus with the power supply on.

Materials

Sperm and eggs of two species of sea urchins (for example, *Strongylocentrotus purpuratus* from the Pacific coast and *Arbacia punctulata* from the Gulf coast)
    18°C incubator
    Clinical centrifuge
    Two multiwell plates (12 wells each)
    Three–five 15-ml screw-top tubes
    Pasteur pipettes and bulbs
    Microscope slides and coverslips
    Hemacytometer
    Compound microscope
    Dissecting microscope
    Seawater (SW), sodium-free seawater (NaFSW), and calcium/magnesium-free seawater (CaMgFSW)
    Diluted india ink
    0.4 mg/ml trypsin inhibitor in SW
    0.5% trypsin in SW
    Calcium ionophore A23187 in SW, NaFSW, and CaMgFSW
    0.65 M NH_4Cl
    Egg jelly from each species

For Part II.G (optional)

    1.5-ml Eppendorf tubes
    Microfuge
    2× SDS–PAGE sample buffer
    Molecular-weight markers
    Microliter pipettor and disposable tips
    Boiling-water bath
    1.5-ml tube support for boiling-water bath
    SDS–PA minigels (precast)
    Electrode buffer

Gel apparatus
Power supply
Coomassie blue staining solution
Destaining solution
Dish for staining and destaining gel

Methods

Use the diagrams of a 12-well multiwell plate (Fig. 1.2) to plan the experiments and record results.

Species 1:_____

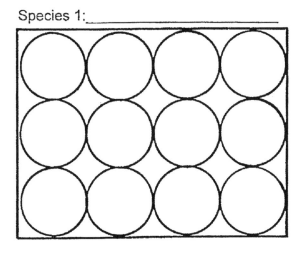

Species 2:_____

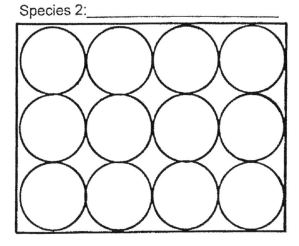

FIGURE 1.2 Twelve-well multiwell plate maps. Use these maps to plan your experiments and to record data.

A. Egg Observations

Suspensions of eggs from two sea urchin species with and without jelly coats are provided.

1. Observe eggs with jelly coats under the microscope. Repeat with dejellied eggs.
2. Mix eggs with jelly coats with a drop of india ink. Repeat with dejellied eggs.
3. Observe the eggs of the other species, repeating the conditions described above.
4. Determine the size of the egg cells and the jelly coat using a hemacytometer.

B. Sperm Observations

Suspensions of sperm from two species are provided.

1. Place a drop of sperm from each species on a hemacytometer and determine their sperm sizes. Note their morphological differences from eggs.
2. Place a drop of sperm on a slide. Add a drop of jelly extract and observe immediately.
3. Repeat with the other species.
4. Place a drop of sperm on a slide. Add a drop of jelly extract from the opposite species and observe immediately.

C. Ion Effects on Gametes and Fertilization

Bottles of SW, NaFSW, and CaMgFSW and sperm and eggs of both species are provided.

1. As a control, first observe normal fertilization in SW by mixing 2 ml egg suspension with 1 ml diluted sperm of the same species in a well of a 12-well multiwell plate. Repeat for the other species. Make a slide of each suspension and observe immediately under the microscope.
2. Use the dissecting microscope to observe and describe the activities of embryos of each species in the multiwell plates at 10- to 15-min intervals throughout the lab period. If desired, make a new slide for higher-resolution observation with the compound microscope.

Eggs and embryos are extremely sensitive to temperature. Eggs and embryos of *S. purpuratus* should be maintained at 18°C in the incubator; the *Arbacia* eggs and embryos should be maintained at room temperature, 25°C (Table 1.1).

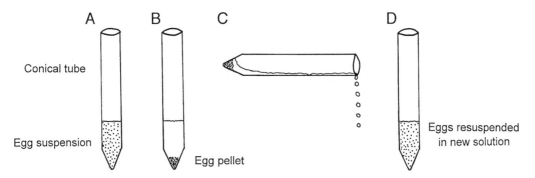

FIGURE 1.3 Transferring eggs to a different seawater solution. A. Transfer 2–3 ml egg suspension to a conical tube. B. Centrifuge at 3/4 top speed for 30 s. C. Decant the seawater. D. Resuspend the egg pellet in a different seawater solution. E. Repeat steps B–D.

Repeat the above procedure in NaFSW and CaMgFSW. To transfer eggs from the stock SW to NaFSW or CaMgFSW (Fig. 1.3), place 2–3 ml of egg suspension in a conical centrifuge tube, place the tube in the clinical centrifuge with a similar tube across from it for balance, and spin at 3/4 top speed for 30 s to pellet the eggs. Gently remove the tube from the centrifuge and decant the seawater. Resuspend the eggs in either NaFSW or CaMgFSW, spin again, decant, and resuspend in fresh SW of the appropriate type. Dilute sperm into each type of SW immediately before use.

1. Observe the gametes before fertilization in NaFSW and CaMgFSW.
2. Observe fertilization in both NaFSW and CaMgFSW.
3. Continue to observe embryos at 10- to 15-min intervals throughout the lab period and as frequently as possible over the next 2 days.

D. Polyspermy

Protease inhibitor (0.4 mg/ml crude soybean trypsin inhibitor in SW) is provided for you.

1. Pellet and resuspend eggs in the protease inhibitor.
2. Incubate 5 min.
3. Add sperm and observe fertilization.
4. Continue your observations at 10- to 15-min intervals throughout the lab period.
5. Label the wells and observe the polyspermic embryos the next day.

E. Cross-Species Fertilization

A solution of 0.5% trypsin (a protease) is provided for your use.

1. Mix sperm from each species with eggs from the other. Observe the fertilization events in both cases.

2. Resuspend a small volume of eggs from each species in 0.5% trypsin solution (*not* the trypsin inhibitor used in II.D, *Polyspermy*) and incubate at room temperature for 30 min.

3. Wash (that is, spin down, decant, and resuspend) twice in SW to remove the trypsin.

4. Repeat cross-species fertilization with the trypsin-treated eggs.

F. Parthenogenetic Activation

Two compounds can parthenogenetically activate eggs: the calcium ionophore A23187, which transports Ca^{2+} across membranes, and NH_4Cl, a weak base that can cross the plasma membrane and cause an increase in intracellular pH. An NH_4Cl solution and solutions of ionophore A23187 in SW, CaMgFSW, and NaFSW are provided.

Ionophore
1. Mix 4 drops A23187 in SW with 4 drops of eggs in SW in a well in your multiwell plate.

2. Repeat the procedure with A23187 in CaMgFSW and NaFSW.

3. Observe and record observations through cleavages for all three samples.

NH_4Cl
4. Add 1 drop NH_4Cl solution to 12 drops of eggs in SW. Observe and record.

5. After 7–10 min, divide the solution into three aliquots. To the first aliquot, add 1 drop diluted sperm; to the second, add 3 drops of A23187 in SW. Add nothing to the third.

6. Record observations through cleavages for all three samples.

G. Gene Expression during Sea Urchin Development (Optional) (See Appendix C)

Prepare Protein Samples
1. Collect samples of sperm, unfertilized eggs, fertilized eggs, two-cell-stage embryos, morulae, ciliated blastulae, gastrulae, and pluteus larvae of each species in separate 1.5-ml Eppendorf tubes. Centrifuge 2 min to pellet the samples. Decant the seawater. Suspend samples in SDS–PAGE sample buffer.

2. Prepare a boiling-water bath and carefully float the protein samples in the boiling water for 5 min. Store the samples at room temperature until all are collected.

3. Reboil the samples for 5 min just before loading them on the gel.

Load Gel and Run

4. Load 10 μl of each protein sample and the molecular-weight marker into consecutive lanes of the SDS–PAGE gel. Note which sample is loaded in each lane.

5. Carefully place the lid on the apparatus and plug the apparatus into the power supply.

6. Electrophorese at 20 mA until the dye in the samples reaches the very bottom of the gel.

7. Turn off the power and unplug the apparatus.

8. Remove the plates from the apparatus and gently separate them.

Visualize Protein Bands

9. Place the gel in a large dish containing Coomassie blue protein stain.

10. Stain the gel for 15–30 min, gently shaking the dish.

11. Pour off the stain and rinse the gel several times in destaining solution.

12. Allow the gel to soak in destaining solution until you can see bands in the gel.

13. Dry the gel on filter paper or plastic sheets.

III. QUESTIONS

A. Egg Observations

1. What are the approximate diameters of the egg cells? Volumes?

2. What are two functions of the jelly coat?

3. What is the average distance between eggs with jelly coats? What is the average distance between eggs that have been dejellied?

4. How close are the suspended ink particles allowed near each type of egg? What is the diameter of the jelly coat?

B. Sperm Observations

1. What are the approximate diameters of the sperm cells? Volumes?

2. Can you see movement or swimming in the sperm suspension?

3. Do both species of sperm exhibit a chemotactic response to the jelly coat extract? Of what benefit is such a chemotactic response to the urchins?

C. Ion Effects on Gametes and Fertilization

Fertilization in SW

1. What is the first evidence you see that fertilization has occurred?
2. How long after the suspensions are mixed does fertilization occur? The first cleavage? The second cleavage?
3. How long does complete formation of the fertilization membrane take? Does the fertilization membrane rise evenly around the egg?

Fertilization in NaFSW and CaMgFSW

4. Describe each gamete in each of the seawater solutions.
5. Is fertilization normal when both egg and sperm are in CaMgFSW? In NaFSW? Give explanations for your observations.

D. Polyspermy

1. How does the fertilization membrane rise around polyspermic eggs? Compare the process to that in normal eggs.
2. Describe cleavage in polyspermic eggs. Is this the result you would expect?
3. Describe the embryos after 24 h.
4. What is the role of the trypsin inhibitor in this instance?

E. Cross-Species Fertilization

Untreated Eggs

1. Describe fertilization in all of the crosses.

Trypsin-Treated Eggs

2. What is the effect of the trypsin on the vitelline membrane? What receptors are located there?
3. Describe the formation of the fertilization membrane after treatment with trypsin.
4. Does normal fertilization occur? How can you tell whether fertilization has occurred in these embryos?

F. Parthenogenetic Activation

1. Summarize and give reasons for your observations of the ionophore-treated gametes.

2. Explain your results from the three aliquots of eggs treated with NH$_4$Cl. Include the mechanism of action of the treatments.

G. Gene Expression during Sea Urchin Development (Optional)

1. Describe the protein patterns on the gels, noting whether bands of common molecular weight occur in the different samples.

2. Determine the molecular weights of the most prominent of the protein bands by graphing them (see Appendix C).

3. Estimate the percentage of the bands that are different and the percentage that are the same in each lane. Drawing on outside resources, can you identify any of the protein bands?

4. What assumptions have we made about the control of gene expression in this exercise?

5. Why is the SDS added to the gel?

6. How is the sieving of the proteins in the gel accomplished?

7. What would happen if you connected the gel apparatus to the power supply with the current reversed?

IV. SUPPLEMENTARY READINGS

Epel, D. (1977). The program of fertilization. *Sci. Am.* 237, 129–138.

Gilbert, S. F. (1994). "Developmental Biology," 4th ed., pp. 121–134, 138–158, 165–170, 202–209, 466–478. Sinauer, Sunderland, MA.

Schoenwolf, G. (1995). "Laboratory Studies of Vertebrate and Invertebrate Embryos," 7th ed., pp. 125–143, 285–296. Prentice-Hall, Englewood Cliffs, NJ.

For Part II.G (Optional)

Gilbert, S. F. (1994). "Developmental Biology," 4th ed., pp. 57–58, 466–478. Sinauer, Sunderland, MA.

Laemmli, U. K. (1970). Cleavage of the structural proteins during the assembly of the head of bacteriophage T4. *Nature* **227**, 680–685.

2 LABORATORY EXERCISE

EARLY AMPHIBIAN DEVELOPMENT

I. INTRODUCTION

Frogs lay large numbers of huge eggs that develop externally in pond water at ambient temperature. Depending on the species, eggs are fertilized just before or immediately after they are laid. Because the frog egg is large enough to be manipulated surgically, a boon to developmental biologists, frogs have been used extensively as a vertebrate embryo model in laboratory studies. Although much of this work was performed in the early 1900s, interest has arisen in reinvestigating the phenomena with molecular-biological techniques in an attempt to improve understanding of the molecular factors involved. In this exercise, you will reproduce some of the most straightforward of these classic experiments.

In this exercise, you will work mainly with the African clawed toad, *Xenopus laevis*. Females of this species ovulate after injection with human chorionic gonadotropin. Each female sheds several thousand eggs. These eggs cleave 1.5 h after fertilization and subsequently at 20- to 30-min intervals. Fertilized and unfertilized eggs will be provided. You may also have the opportunity to work with fertilized eggs of the axolotl, *Ambystoma mexicanum,* imported from the axolotl colony at Indiana University.

Experimental Rationale

In Part II.A, you will examine the general physical characteristics (morphology, size, and volume) of frog gametes. You will also observe the characteristics of early cleavage in frog embryos, for comparison with cleavage of other species you encounter in this laboratory.

Recent work has yielded a greater understanding of the early inductive interactions that organize the embryonic axes. In Parts II.B and II.C, you will examine the role of cortical rotation in establishing the early embryonic axes.

In Parts II.D and II.E, you will probe the organization of frog egg cytoplasm by examining the capacity of separated blastomeres to form complete embryos and the effects on normal embryo formation of reorienting the early cleavage planes.

Developmental biologists and geneticists are fascinated with the contributions of the male pronucleus to both early development and later life. In Part II.F, you will evaluate several means of artificially activating frog eggs and examine the male contributions to early frog development.

Parts II.G and II.H of this exercise should lead you to develop a three-dimensional mental image of how the early frog organizes itself into a multilayered embryo.

In Part II.I, you will observe normal development of axolotl embryos and try your skills at early embryo dissection using these large and stunningly beautiful embryos.

II. EXPERIMENTAL PROTOCOL

Materials

Unfertilized and fertilized amphibian eggs
Sperm suspension
Axolotl embryos (optional)
Two pairs of fine forceps, scalpel
35- and 60-mm petri dishes
Small pieces of filter paper
Quartz dish or cuvette
Hand-held UV light source
Dissecting microscope
Compound microscope
Capillary tubes for making glass needles
Bunsen burners
Solution of diluted frog blood
Solution of 1 mM $CaCl_2$

Red, green, blue, and yellow modeling clay
Prepared slides of frog gastrula, early and late neural tube, early and late neural groove, neural fold, neural plate, and 4-mm embryos
Pasteur pipettes and bulbs
8 mM DTT solution for dejellying eggs
100% Steinberg's solution
70% Steinberg's solution
20% Steinberg's solution

Methods

Work individually and compare your observations and results with those of others. Save remaining normal tadpoles for use in Laboratory Exercise 5, *Amphibian Metamorphosis.*

A. Observations of Gametes and Cleaving Embryos

1. Observe fertilized eggs under the dissecting microscope throughout the lab period, at least to the 8-cell stage (Table 2.1).
2. Determine the sizes of frog sperm, eggs, and 2-, 4-, and 8-cell embryos.

TABLE 2.1

Timetable for Normal Embryonic Development of *Xenopus laevis*

	Time after fertilization	
Stage	at 18°C	at 22–25°C
1-cell, unfertilized	0 h	0 h
2-cell	1.5 h	1.5 h
4-cell	2.5 h	2 h
8-cell	3.5 h	2.25 h
16-cell, early cleavage	4 h	2.75 h
32-cell, mid-cleavage	5 h	3 h
Blastula	6 h	5 h
Gastrula	8 h	9 h
Yolk plug	15 h	13.5 h
Neural plate	18 h	16 h
Neural fold	21 h	18 h
Neural groove	25 h	20 h
Neural tube	30 h	21 h
Hatching	72 h	50 h
Feeding tadpoles	5 days	98 h
Hind limb stage	38 days	24 days

B. Embryo Inversion

1. Within 15 min after fertilization, transfer several fertilized eggs to a small filter-paper square with a minimum amount of water. Orient the eggs so that the pigmented animal pole is up.

2. Use a scalpel to spread the jelly coat of each egg down to the filter paper, so that the jelly holds the egg to the filter paper. Allow the jelly to dry slightly.

3. Invert the paper and embryo in a petri dish of well water or 20% Steinberg's solution.

4. After a few minutes, hold the dish up and examine the embryos through the bottom of the dish, checking to see that the dark animal pole of the embryo is pointed downward in the solution.

5. Keep the embryos inverted through the 2-cell stage, and then turn them back over.

6. Observe the embryos at the neural plate/neural tube and tailbud stages, looking for the presence of two neural plates/tubes.

C. Prevention of Cortical Rotation by UV Irradiation

1. Place five unfertilized eggs into each of five petri dishes.

2. Fertilize the eggs with diluted frog-sperm suspension. Note the time of fertilization. Wash the eggs to remove excess sperm.

3. Within 20 min after fertilization, transfer the embryos to the quartz dish or cuvette. Irradiate the eggs from the bottom using a hand-held UV light source for 5, 15, 30, and 60 s and 5 min. After irradiation, return the embryos to the petri dishes to continue development.

4. Observe the embryos over the next several minutes, hours, and days. Record the effects of UV treatment on subsequent embryonic development.

D. Blastomere Separation

1. Place a 2-cell embryo on a filter-paper square and remove the jelly coat with a scalpel and fine forceps. Do not allow the dejellied embryo to dry. Alternatively, eggs can be dejellied by a brief (less than 2-min) treatment with 8 mM DTT dejellying solution. After treatment, wash the embryos 3–4 times with 100% Steinberg's solution, then with 70% Steinberg's solution, and gradually dilute the Steinberg's solution to 20% over the next hour. Dejellied eggs and embryos are very fragile, so use care in handling them.

2. Using fine forceps, carefully remove the fertilization membrane from the embryo.

3. After dejellying and demembranating the embryos, separate the blastomeres by running a fine glass needle down the midline between them.

4. Place the separated blastomeres in individual containers. If filter paper was used, submerge the paper and carefully tease the blastomeres off the paper.

5. Cover with well water or 20% Steinberg's solution.

6. Observe the embryos during subsequent developmental stages.

7. Additional optional experiments would be to ablate (destroy but leave attached) one of the two blastomeres and to duplicate Spemann's hair-loop experiment on eggs approaching the first cleavage.

E. Cleavage Plane Reorientation

1. Place a 4-cell embryo in well water or 20% Steinberg's solution in the inverted lid of a petri dish (see Fig. 2.1).

2. Place the bottom of the petri dish, open side up, over the embryo and fill it with water until the embryo appears slightly squashed.

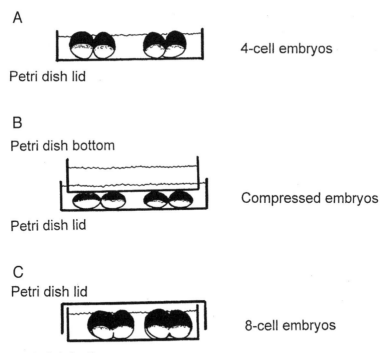

A

Petri dish lid

4-cell embryos

B

Petri dish bottom

Petri dish lid

Compressed embryos

C

Petri dish lid

Petri dish bottom

8-cell embryos

FIGURE 2.1 Reorientation of the third cleavage plane. Place 4-cell embryos in the inverted lid of a 60-mm petri dish and cover with water. Place the bottom of the petri dish inside the inverted lid over the embryos. Add water by drops to the petri dish bottom until the embryos are slightly compressed. After the embryos progress to the 8-cell stage, lift off the petri dish bottom, transfer the embryos to the petri dish bottom, add well water or 20% Steinberg's solution, and cover with the lid. Record observations of the embryos over the next several days.

3. Observe the embryo during the next cleavage using a dissecting microscope, noting whether the cleavage is meridional or equatorial.

4. After this cleavage, release the pressure on the egg by removing the petri dish bottom. Transfer the embryo in the bottom of the petri dish and cover it with well water or 20% Steinberg's solution.

5. Observe the embryo after several hours and days, and record the effects of this treatment on subsequent development.

F. Parthenogenetic Activation

This section of the exercise should be done early in the lab meeting, so that sufficient time remains to observe whether cleavage occurs.

1. Three methods of parthenogenetic activation are available for this part of the exercise: (a) pricking unfertilized eggs in a solution of frog blood, (b) pricking unfertilized eggs in a solution of 1 mM CaCl2, and (c) inseminating unfertilized eggs with UV-irradiated sperm.

a. To induce parthenogenesis with frog blood, place five unfertilized eggs on a dry slide or petri dish and cover with a small amount of frog blood. Gently and firmly, prick each egg with the sharp point of a glass needle. The puncture should be in the animal hemisphere, but not at the center where the second maturation (meiotic) spindle is located (possibly seen as a clearing in the black pigment of the animal cap). A small amount of egg cytoplasm may leak out at the site of puncture. Cover the eggs with well water or 20% Steinberg's solution, wash off the blood, and refill the dish with well water or 20% Steinberg's solution.

b. To induce parthenogenesis with 1 mM $CaCl_2$, follow instructions in (a), substituting $CaCl_2$ for frog blood.

c. To induce parthenogenesis using UV-irradiated sperm, place 1 ml of sperm suspension in a 35-mm petri dish positioned under a UV light source and irradiate the sperm suspension for 15 min. Swirl the dish occasionally during exposure. Place five unfertilized eggs in a petri dish and fertilize them with the irradiated sperm. *Caution:* Do not expose your skin or eyes to UV irradiation unnecessarily and do not look directly at the lamp.

2. Allow embryos to develop and record your observations.

G. Modeling Early Embryonic Development

1. Produce at least one three-dimensional embryo from each of the following groups: Group I: early gastrula, late gastrula. Group II: early neurula (neural plate), mid neurula (neural fold).

2. Use yellow modeling clay to represent endoderm, red for mesoderm, blue for ectoderm, and green for chordamesoderm (notochord). Serially section your models and compare them with the embryo sections on the prepared slides and with the micrographs in your manuals.

3. Gain a working knowledge of the location of each germ layer at specific times of development.

H. Gastrula and Neurula Serial Sections

1. Study an embryological atlas (e.g., Schoenwolf, 1995) and/or examine slides of early frog embryos in the slide boxes.

2. Obtain a working knowledge of where each germ layer is located at the following specific developmental times: gastrula, early and late neural tube, early and late neural groove, neural fold and plate, 4-mm embryo.

I. Observations of Axolotl Embryos (Optional)

1. Examine the axolotl embryos provided.

2. Observe subsequent development over the next several days and record your observations.

3. If feasible, perform the following experiment:

a. Dejelly eggs with 8 mM DTT dejellying solution and remove the fertilization membrane with fine forceps as described above for frog eggs.

b. Using fine forceps or fine glass needles, cut the upper animal cap free from the vegetal half of several of the embryos. Place the separated halves in different dishes.

c. Observe development of the separated halves of the embryos over the next hours and days.

III. QUESTIONS

A. Observations of Gametes and Cleaving Embryos

1. Compare the sizes of frog sperm and eggs.
2. Can you see the gray crescent?
3. Describe the cleavage furrow.
4. What is the orientation of each of the first three cleavage furrows?
5. Is the embryo changing size during this period?

B. Embryo Inversion

    1. Summarize and give reasons for your observations.

C. Prevention of Cortical Rotation

    1. Summarize your results and compare these eggs with unirradiated eggs.

D. Blastomere Separation

    1. Summarize and give reasons for your observations.

E. Cleavage Plane Reorientation

    1. Does the embryo develop normally?
    2. What does this result indicate about mosaic and regulative development?

F. Parthenogenetic Activation

    1. Summarize your results and compare these eggs with normally inseminated eggs.

G. Observations of Axolotl Embryos

    1. Compare the appearance of axolotl embryos with that of *Xenopus* embryos.
    2. At what stage of development did you begin your observations?
    3. Do the separated halves of the embryo develop into recognizable tadpoles?

IV. SUPPLEMENTARY READINGS

Gilbert, S. F. (1994). "Developmental Biology,"4th ed., pp. 155–158, 170–171, 211–222, 244–252, (optional) 586–620. Sinauer, Sunderland, MA.

Kirschner, M. W., and J. C. Gerhart (1981). Spatial and temporal changes in the amphibian egg. *BioScience* 31, 381–388.

Schoenwolf, G. (1995). "Laboratory Studies of Vertebrate and Invertebrate Embryos,"7th ed. pp. 5–12, 21–26. Prentice-Hall, Englewood Cliffs, NJ.

Rugh, R. (1948). "Experimental Embryology: A Manual of Techniques and Procedures,"pp. 66–68, 77–81, 116–121, 183–196. Burgess, Minneapolis, MN.

3 LABORATORY EXERCISE

EMBRYONIC CHICK DEVELOPMENT

I. INTRODUCTION

Embryos of the chicken, *Gallus domesticus,* have served for many years as an excellent system for studying higher vertebrate development. Fertilization in chickens is internal, so the embryo in a newly laid hen's egg has already undergone the first stages of cleavage. Once outside the hen, embryo development stops at the embryonic-shield stage until the temperature is raised to 37°C, at which time development continues.

Chick organ development is similar to that in mammals, but the chick embryo is much more amenable to experimental study of organogenesis than is a human embryo. Chick embryos develop quickly, hatching after 21 days of incubation. In Fig. 3.1, notice the internal features of a fertilized, unincubated egg and an egg after 4–5 days of incubation.

Experimental Rationale

In the first of this series of experiments you will observe normal chick embryo development by examining prepared slides of chick embryos of several developmental stages in Part II.A. To complement your understanding of chick

A

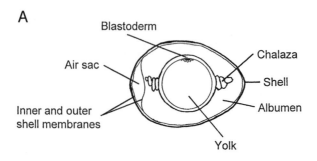

B

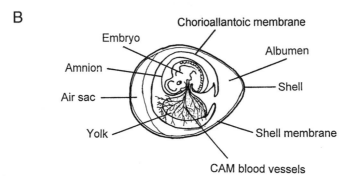

FIGURE 3.1 Internal features of fertilized chick eggs. A. Features of an unincubated egg. B. Features of an egg incubated for 4–5 days.

development from the slides, you will also window an egg and observe it over the next 2 weeks for Part II.B.

In Part II.C, you will determine whether a chick embryo can regenerate a new limb if the limb bud is removed very early in development.

In Part II.D, to learn about limb and/or eye development in the chick embryo, you will perform chorioallantoic grafts.

In Part II.E, you will gain insight into the cellular movements and the structures formed in various stages of chick development by staining embryos with vital dyes.

In Part II.F, you will perform experiments on explanted embryos and study heart development in the early chick embryo.

II. EXPERIMENTAL PROTOCOL

CAUTION: The Howard–Ringer's (H–R) solution used in these experiments contains penicillin and streptomycin. Please tell the instructor if you are allergic to antibiotics.

Materials

Prepared slides of whole mounts and serial sections of 18-, 24-, 33-, 48-, and 72-h chick embryos

Fertilized eggs, incubated for the specified length of time

Dissecting microscopes

Rulers

Pencils

Dissecting needles, forceps, and capillary tubes for making fine needles

Bunsen burner and striker

Small files

Melted wax

Kimwipes

22 × 22 mm coverslips

15 and 37°C incubators

70% ethanol

Sterile H–R solution

Sterile Pasteur pipettes and bulbs

Organ-culture plates containing agar–albumin

Methods

A. Early Chick Development

1. Study the prepared slides of the 18-h whole mounts and the whole mounts and serial sections of 24-, 33-, 48-, and 72-h embryos for your study of early chick development.

2. Determine the sizes of the different ages of embryos. Pay particular attention to heart development.

B. Egg Windowing (Fig. 3.2)

1. Using a pencil, place an "X" on the top side of an egg in the 15°C incubator. Remove the egg from the incubator, and maintain this orientation as you carry the egg back to your lab bench.

2. Mark a 1.5- to 2-cm square around the "X" with a pencil (the solvents in permanent markers are toxic to the embryos). Wipe the outside of the egg and your dissection instruments with 70% ethanol to sterilize them.

3. Punch a small hole in the broad end of the shell with your dissecting needle to allow air to escape from the air sac.

4. Carefully saw through the shell on the pencil lines with the file provided. Remove the shell square, leaving the shell membrane intact. Clear

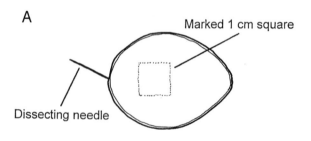

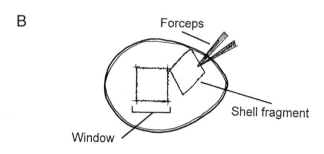

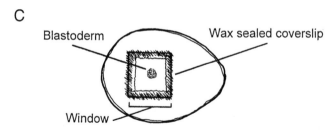

FIGURE 3.2 Windowing a chick egg. A. Use a pencil to mark a 1-cm square on the egg surface. With a dissecting needle, drill a hole in the blunt end of the egg to vent the air sac. B. Use a small file to saw through the egg shell, leaving the shell membranes intact. Wipe the dust and shell chips from the egg surface. Use forceps to peel back, gently, the 1-cm-square egg-shell fragment. Then, gently peel back the shell membranes to expose the embryonic blastoderm. C. Place a 22-cm-square coverslip over the hole, and seal around the glass with melted wax.

away any dust and debris by wiping lightly with 70% ethanol before proceeding to the next step.

5. Peel away the shell membrane so that you can see the developing embryo. Note: If the egg cracks slightly at any time while you are opening the window, you will need to obtain a fresh egg and start the process over.

6. Use one of the following three methods for sealing the window shut.

   a. Place a coverslip over the square hole and seal around it completely with melted wax (a permanent closure; may fog).

b. Seal the hole with parafilm (a temporary closure; potential contamination).

c. Cover the hole with sticky tape (a permanent closure; may fog).

7. Place the egg in the 37°C incubator and *view it daily* over the next 2 weeks.

C. Limb Regeneration

1. Window an egg that has been incubated for 3–4 days, remembering to sterilize the egg and your instruments and to vent the air sac first.

2. Carefully cut a flap in the vitelline membrane to expose the embryo's wing bud or hind limb bud area. Then make a small hole in the amnion.

3. Insert fine forceps or a fine glass needle through the hole into the amnion and pinch or cut off a wing bud without destroying the underlying tissue (see Fig. 3.3A). Remove the bud with a sterile pipette or leave it floating in the amnion. If you leave it in place, be sure to push it far away from its former site.

4. Seal the window with a coverslip and wax or any of the other methods, and allow the embryo to develop for 5–7 days to determine whether the limb is able to regenerate.

5. *View it daily* and record your results.

D. Tissue Grafting

Host Preparation

1. Candle an 11- to 12-day egg and mark on the shell the location of a major blood vessel bifurcation far from the developing embryo.

2. Sterilize the egg and your instruments as before, and window the egg to expose the bifurcated vessel.

3. Cover the window temporarily with parafilm and return the egg to the 37°C incubator or put it under a lamp to keep it warm.

Graft Preparation

4. Prepare the graft tissue by windowing a 4- to 5-day egg to expose the embryo. Cut a larger than normal window to make working on the embryo easier. Decide which tissue to graft.

5. Remove either an eye, a patch of dorsal skin, a small piece of bone, or the limb bud (including some somitic tissue underlying the limb) from the 3- to 4-day embryo by either:

a. cutting out and transferring the blastoderm to a petri dish containing sterile Howard–Ringer's solution and removing the limb bud and some adjacent somites (see *Chick Embryo Explantation,* below) or

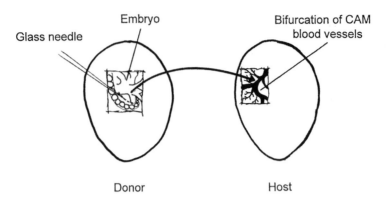

FIGURE 3.3 Tissue graft to blood vessels of the chorioallantoic membrane. Window the donor egg. Dissect the graft tissue free from the donor embryo. Candle the host egg and prepare a window over a bifurcation in a blood vessel of the chorioallantoic membrane. Gently scrape the vessel, place the graft tissue at the scraped area, and seal the window with a glass coverslip.

    b. carefully removing the limb bud and surrounding somites from the embryo in the egg and storing it in a sterile dish of sterile H–R solution until the host is ready.

Graft Transfer to Host

    6. Remove the parafilm cover from the host and place the eye or limb bud near the branch point of the bifurcating blood vessel (see Fig. 3.3).

    7. Gently scratch the vessel in the chorioallantoic membrane with the dissecting needle without rupturing it. A slight amount of bleeding is optimal; too much bleeding is lethal.

    8. Place the graft tissue over the scratched area so that the limb bud or eye is up.

    9. Remove any excess saline with a pipette so that the graft will not float away.

    10. Seal the window with a glass coverslip and return the egg to the incubator. *View it daily* over the next several days. At the end of 1 week, open the egg shell, examine the grafted tissue, and record results.

E. Programmed Cell Death

    1. Window a 4- to 5-day egg and remove approximately two pipettes full of albumin. The window should be large enough to allow clear viewing of the embryo.

    2. Remove the vitelline membrane and poke a hole in the amnion.

    3. Using a pipette drawn out over a flame, add one drop of the vital dye 0.4% Trypan blue to the embryo.

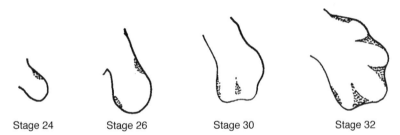

Stage 24   Stage 26   Stage 30   Stage 32

FIGURE 3.4 Regions of cell death in the developing chick limb. Numbers refer to stages of the Hamburger-Hamilton development series (V. Hamburger and H. L. Hamilton, *J. Morphol.* 88, 49–92, 1951.)

4. Wait 2–3 min until the dead cells begin to take up the stain. Continue adding stain until the areas of cell death are clearly visible. Compare your results with diagrams in Fig. 3.4.

F. Induction of Cardia Bifida

Cardia bifida induction requires that you first remove or explant the developing embryo from the egg for surgery. Proficiency in the explantation technique requires patience and is critical for successful cardia bifida formation. Two alternative methods for explantation are described in detail and illustrated below in Fig. 3.5.

1. Eggs have been incubated for you so that the embryos are at the 27-h stage, having 6–7 somites.
2. Use one or more eggs for practice. Ultimately, explant two blastoderms for the experiment.
3. Prepare one or two glass needles for the surgery. To maintain humidity in each organ culture chamber, place a Kimwipe moistened with H–R solution around the inside perimeter of the dish.
4. Perform the cardia bifida surgery on one of the two explanted blastoderms (see Fig. 3.6). After the explanted embryo has rested for 20 min, make a cut through the middle of the anterior intestinal portal, by passing the needle several times through the floor of the foregut. Refer to Fig. 3.6 to determine where to operate on the ventral side of the explanted embryo.
5. Return the embryo to the incubator and *view it after 24 h.*

Explantation Method A

1. Cut several doughnut-shaped discs of filter paper 1 in. in outside diameter and 0.75 in. inside diameter.

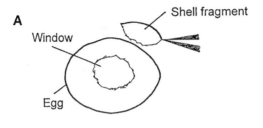

A

Shell fragment

Window

Egg

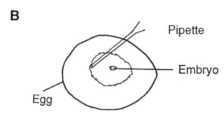

B

Pipette

Embryo

Egg

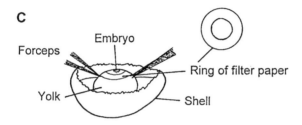

C

Embryo

Forceps

Ring of filter paper

Yolk

Shell

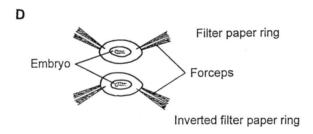

D

Filter paper ring

Embryo

Forceps

Inverted filter paper ring

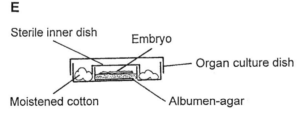

E

Sterile inner dish

Embryo

Organ culture dish

Moistened cotton

Albumen-agar

FIGURE 3.5 Chick embryo explantation. A. Prepare a window on an egg, but do not seal it closed. B. Withdraw albumen from the egg with a sterile Pasteur pipette so that the yolk is exposed. C. Remove more of the shell so that only half remains. Cut a ring of filter paper with inner diameter slightly smaller than the sinus terminalis, which contains the embryo. Carefully center the ring over the edges of the sinus terminalis. Grasp the ring with forceps and cut around the ring completely to free the embryo from the membrane. D. Using two forceps, slowly lift the ring and embryo from the yolk. Invert it so that the embryo's ventral side is up, if the embryo is explanted for use in Part II.F, *Induction of Cardia Bifida*. E. Place the ring and embryo on albumen–agar in the inner chamber of an organ culture dish. Add sterile Howard–Ringer's solution.

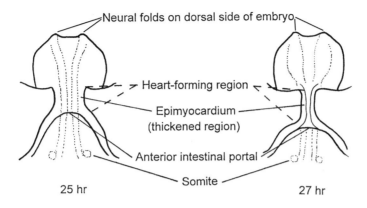

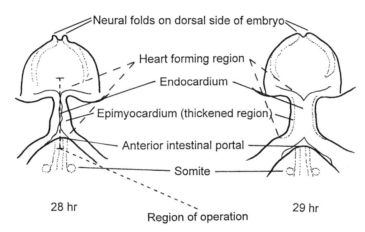

FIGURE 3.6 Induction of cardia bifida. Ventral view of chick development showing fusion of the epimyocardial folds (at ~28 h of incubation) during heart formation. To prevent fusion of the folds and induce cardia bifida, use a glass knife to cut the along the region of operation in an embryo incubated for 28 h.

2. Gently crack an egg and pour the embryo, yolk, and albumin into a dish containing warm H–R solution. Be careful not to let the shell tear the vitelline membrane.

3. Lightly scrape the surface of the egg over the blastoderm until it appears almost dry and place one of the filter paper circles on the scraped surface with the blastoderm centered in the middle of the hole. Make sure the ring makes complete contact with the surface of the yolk.

4. Hold the ring with a pair of blunt forceps and cut through the vitelline membrane around the outside of the paper ring. Use a dissecting needle to ensure that the vitelline membrane is completely cut around the ring.

5. Using two pairs of forceps, lift the ring with the blastoderm attached away from the yolk, turn it over, and transfer it to an organ-culture dish. The embryo

should be on top of the paper ring, ventral-side up. Cover and place the embryo in the incubator to recover for about 20 min before continuing the operation.

Explantation Method B

1. Gently crack an egg and pour the embryo, yolk, and albumin into a dish containing warm H–R solution. Be careful not to let the shell tear the vitelline membrane.

2. Grasp the chalaza with the forceps and cut through the vitelline membrane, making a circle around the blastoderm approximately 0.25 in. from its margin. By cutting in one direction and moving the yolk mass in the other direction, you can quickly excise the blastoderm.

3. Grasp the margin of the blastoderm, and roll it back away from the yolk.

4. Transfer the blastoderm to the organ-culture chamber and orient it with the ventral side of the embryo up.

5. Cover and place the embryo in the incubator to recover for about 20 min before continuing with the operation.

Explantation Method C

1. Punch a hole in the air sac of the egg.

2. Stabilize the egg shell in a "nest" of Kimwipes in a petri dish or part of an egg carton.

3. Make a small opening in the top of the egg over the embryo, as if beginning to make a window, but continue removing small pieces of shell, until approximately half the shell is removed and the embryo is easily accessible, as in Fig. 3.5.

4. Explant the embryo either with or without the aid of a paper doughnut, as described in Method A or Method B.

III. QUESTIONS

A. Early Chick Development

1. Produce a graph of length versus time for the 18-, 24-, 33-, 48-, and 72-h embryos.

2. Does the length of the embryo increase at a constant rate?

3. Would your graph of chick embryo development be similar to one produced for the frog embryo? the sea urchin embryo? the mammalian embryo?

B. Egg Windowing

1. Make drawings of the embryo over a 7- to 14-day period in your notebook. Make note of any changes during the chick's development.

C. Limb Regeneration

1. Make drawings of the embryo over the 5- to 7-day period. Make note of any changes in the limb bud area during the chick's development.

D. Tissue Grafting

1. Make drawings of the embryo over the next few days. Make note of any changes in the graft or area surrounding the graft.
2. Which eggs are used as controls? Why?

E. Programmed Cell Death

1. Sketch the areas of cell death. Are the areas of degeneration related to the shape of the embryo?

F. Induction of Cardia Bifida

1. What is the sequence of layers you look through to see the embryo from above? From below?

IV. SUPPLEMENTARY READINGS

DeHaan, R. L. (1959). Cardia bifida and the development of pacemaker function in the early chick heart. *Dev. Biol.* 1, 586–602.

Gilbert, S. F. (1994). "Developmental Biology," 4th ed., pp. 185–187, 223–234, 248–252, 267–270, 324–332, 342–350. Sinauer, Sunderland, MA.

Saunders, J. W., Jr. (1966). Death in embryonic systems. *Science* 154, 604–612.

Schoenwolf, G. (1995). "Laboratory Studies of Vertebrate and Invertebrate Embryos," 7th ed., pp. 27–30, 34–57, 63, 67–79, 82–98. Prentice-Hall, Englewood Cliffs, NJ.

Tickle, C. (1981). Limb regeneration. *Am. Sci.* 69, 641–646.

4 LABORATORY EXERCISE

DROSOPHILA GENE EXPRESSION

I. INTRODUCTION

Studies of the fruit fly *Drosophila* have generated a wealth of genetic information and recently have begun to contribute to our understanding of such fundamental problems in developmental biology as axis determination, pattern formation, and regulation of gene expression. Several features make this organism a good research tool for genetic and developmental studies.

One attractive feature is the rapid life cycle of *Drosophila*, which is completed in about 2 weeks at 25°C. Fertilized eggs hatch into first-instar larvae within 1 day. Over the next 4–5 days, the larvae molt twice to become third-instar larvae. Late third-instar larvae crawl away from their food onto a dry surface and form pupae. After 4–6 days, each pupa ecloses into an adult fly, which is ready to lay eggs within 1 day of hatching.

An additional attractive feature of this research organism is the polytene chromosomes found in nuclei of larval salivary-gland cells and cells of certain other larval tissues. Each chromosome in the salivary gland nucleus replicates itself ~1000 times, and the copies remain together, lined up parallel to each other. These chromosomes are easily visible by light microscopy. Small differences in chromosome density along the chromosome cause the appearance of characteristic bands on each chromosome. When a particular gene is being transcribed, the DNA in that region becomes less compacted, or "puffed"; such "puffed" regions regress when the gene is no longer being transcribed. During larval development, specific regions (corresponding to genetic loci) along the polytene chromosomes are always puffed at a certain time, and different regions are puffed at different times. Thus, studies of polytene chromosome puffing patterns can be used to monitor changes in gene activity during larval development.

Changes in polytene chromosome puffing patterns also can be induced by environmental stress, such as heat shock. If *Drosophila* is subjected to a brief heat treatment (30 min at 37°C), puffs that were active at the time of the heat treatment regress, and new puffs appear at nine sites along the chromosomes (the heat-shock loci). In *Drosophila melanogaster*, these sites are at bands 33B, 64C, 64F, 67B, 70A, 87A, 87C, 93D, and 95D. Synthesis of most cellular proteins halts, but synthesis of heat-shock proteins (HSPs) is induced. The mechanism used to regulate activity of the heat-shock genes is thought to be similar to a general mechanism used for expression of all genes (even those expressed during development), so the heat-shock response serves as a model system for study of gene regulation.

Experimental Rationale

In Part II.A of this exercise, you will examine the exterior anatomy of eggs, larvae, pupae, and anesthetized adult flies.

In Part II.B, you will gain an appreciation of one feature that makes fruit flies popular as the subject of genetic research and studies of gene expression. You will examine polytene chromosome puffing patterns in salivary glands from normal larvae. You will next compare polytene chromosome puffing patterns in salivary glands from control larvae and from larvae that were given a heat-shock treatment.

In Part II.C, you will compare the proteins present in control and heat-shocked larvae and larval salivary glands using SDS–PA gel electrophoresis, to complement your studies of larval polytene chromosomes.

II. EXPERIMENTAL PROTOCOL

CAUTION: Ether, used for anesthetizing adult fruit flies, is extremely flammable.

Materials

Vials of *Drosophila* cultures
Dissecting microscope
Compound microscope
Two pairs of fine-point forceps or one pair of forceps and one dissecting
needle
Kimwipes
Depression slides, if available
Microscope slides and coverslips
Nail polish
1.5-ml Eppendorf tubes
Microfuge
2 × SDS–PAGE sample buffer
Molecular-weight markers
Microliter pipettor and disposable tips
Boiling-water bath
1.5-ml tube support for boiling water bath
SDS–PA minigels (precast)
Electrode buffer
Gel apparatus
Power supply
Coomassie blue staining solution
Destaining solution
Dish for staining and destaining gel

Methods

A. Examination of Eggs, Larvae, Pupae, and Adults

1. Examine a bottle containing a week-old culture of *Drosophila* from which the adult flies have been removed.
2. Find examples of eggs, larvae (there are three molting stages or instars), and pupae. Examine the anesthetized adult flies that were removed from the bottle, which are available separately (see Fig. 4.1).
3. Remove samples of each stage and examine them with a dissecting microscope.

B. Polytene Chromosome Squashes

Squashes of Salivary Glands from Control Larvae
1. Select a large, third-instar larva that has started to crawl up the wall of the culture bottle, and place it on a depression slide in a drop of insect saline.

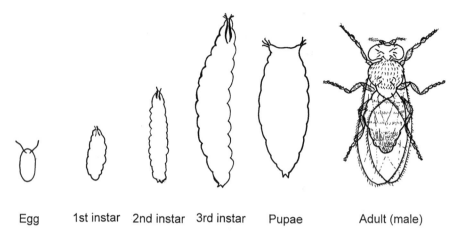

Egg   1st instar   2nd instar   3rd instar   Pupae   Adult (male)

FIGURE 4.1 Developmental stages of *Drosophila*. Not drawn to scale.

2. While viewing through a dissecting microscope, grasp the head of the larva with one pair of forceps and, with another pair, gently pull the body away from the head. Alternatively, a dissecting needle can be used to immobilize the head while a pair of forceps is used to grasp and pull the body away from the head. The salivary glands will trail immediately behind the head (along with many other tissues) and may have yellowish fat bodies attached along the outer edge (see Fig. 4.2). Keep the glands moist at all times and dissect them free of other larval tissues.

3. Fill a well on the depression slide with fixative. Still using the dissecting microscope, transfer the glands to the fixative well on the same slide for ~1 min. Try not to release the glands in the fixative.

4. Finally, transfer the glands to a small drop of 2% aceto orcein stain on a regular microscope slide and leave them to stain for 10 min. Do not let the glands dry out during this time—the stain is volatile.

5. Gently place a coverslip over the stained glands. Put a blotter of folded Kimwipe over the coverslip. Place your thumb on the blotter, and apply firm pressure with your thumb. Be careful not to let the slide move relative to the coverslip.

6. Seal the edges of the coverslip with nail polish to prevent drying of the preparation.

7. First examine your chromosome preparation using 10× power on your compound microscope, then switch to 40× power to see individually spread chromosomes. With luck, your chromosome preparation will look like Fig. 4.3.

Squashes of Salivary Glands from Heat-Shocked Larvae

8. Prepare a polytene chromosome squash as described above, from larvae that have been heat shocked by incubation at 37°C for 30 min.

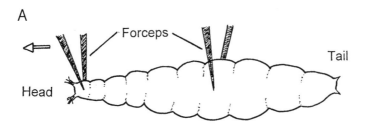

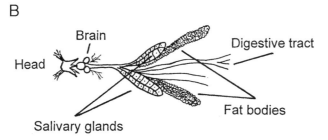

FIGURE 4.2 Dissection of salivary glands from a *Drosophila* third-instar larva. A. Remove a large third-instar larva from the side of the culture vessel near the food. Submerge it in a pool of saline solution. Grasp the midbody of the larva and hold it stationary with one pair of forceps or an insect pin. Grasp the head of the larva firmly with a pair of forceps, and pull to separate the head from the body. B. Locate the salivary glands, which trail behind the brain and head cuticle. Dissect the glands free of the attached fat bodies and transfer them to a pool of fixative, then to aceto orcein stain.

Hints

9. It will take practice to isolate the glands and also to spread the chromosomes sufficiently to see individual chromosomes. While one set of glands is staining, begin isolation of more glands.

10. If the chromosomes are not sufficiently stained, make another preparation, and stain it for a longer time (for example, up to 30 min).

11. If the chromosomes are in a tight ball and not spread out enough to reveal the individual chromosomes, apply more pressure when making the squashes. Also, try rolling your thumb back and forth over the coverslip, again being careful not to let the slide move relative to the coverslip.

C. SDS–PA Gel Electrophoresis

1. Collect five pairs of polytene chromosomes from control and heat-shocked larvae. Place each set in an individual 1.5-ml plastic tube and dissolve the glands in 1 vol of 2× sample buffer for SDS–PAGE.

2. Collect a single control larva and a heat-shocked larva, place each of them in an individual 1.5-ml plastic tube, and dissolve them in ~25 μl of 2× sample buffer.

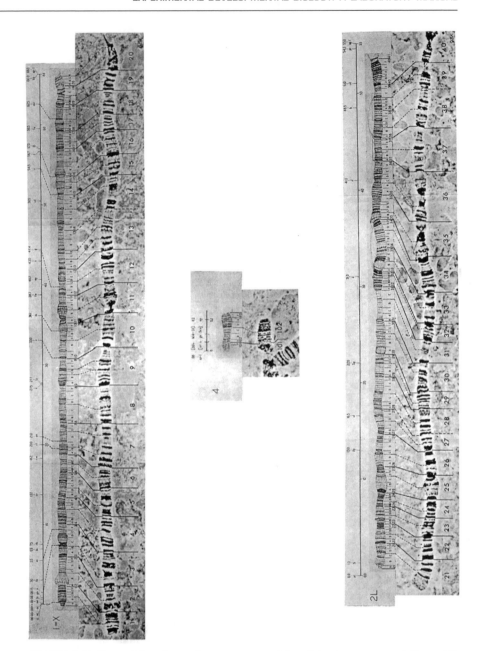

FIGURE 4.3 *Drosophila* polytene chromosome map. Map of the chromosomes of *Drosophila* showing the distinct banding patterns and loci. Reproduced from the original by C. B. Bridges (*J. Hered.* **26**, 60, 1935) with permission of the copyright holder.

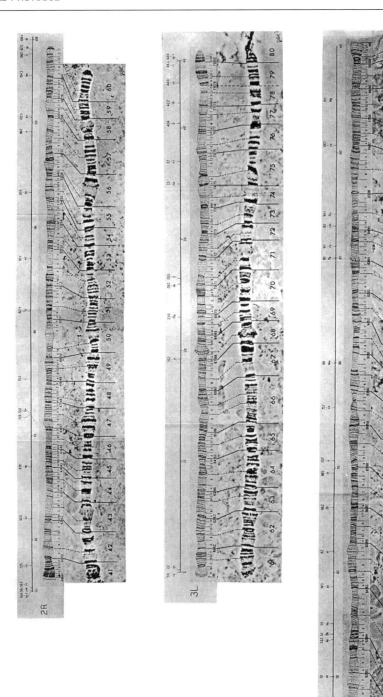

FIGURE 4.3 *(Continued)*

3. Mince and dissolve the samples by pipetting. Boil the samples and load them onto a precast SDS–PA gel in the following order: a. control salivary glands from non-heat-shocked larvae; b. salivary glands from heat-shocked larvae; c. control, whole third-instar larvae; d. whole heat-shocked larvae.

4. Electrophorese as described in Appendix C.

III. QUESTIONS

A. Examination of Eggs, Larvae, Pupae, and Adults

1. What is the size difference between an egg and an adult *Drosophila?*
2. How much do larvae grow between the first and third larval molts?

B. Polytene Chromosome Squashes

1. Identify the individual *Drosophila* polytene chromosomes using the map in Fig. 4.3.

2. Using the map in Fig. 4.3, identify loci that are puffed in control larvae but have regressed in heat-shocked larvae.

3. Identify loci that are puffed in heat-shocked larvae but not in control larvae.

C. SDS–PA Gel Electrophoresis

1. Describe the protein patterns on the gels, noting whether there are any bands of common molecular weights in the different samples.

2. Describe any differences in the protein bands present on gels of control and heat-shocked samples.

3. Describe any differences between the protein bands present in salivary glands and bands present in the whole-larva samples.

IV. SUPPLEMENTARY READINGS

Ashburner, M., and Bonner, J. J. (1979). The induction of gene activity in *Drosophila* by heat shock. *Cell* 17, 241–254.

Gilbert, S. F. (1994). "Developmental Biology," 4th ed., pp. 48–53. Sinauer, Sunderland, MA.

Ritossa, F. M. (1964). Experimental activation of specific loci in polytene chromosomes of *Drosophila. Exp. Cell Res.* 35, 601–607.

5 LABORATORY EXERCISE

AMPHIBIAN METAMORPHOSIS

I. INTRODUCTION

During metamorphosis, dramatic morphological and physiological changes occur that prepare the amphibian tadpole for its transition from aquatic to terrestrial life. Tadpoles grow limbs and lose their tails; their mouths and guts change to accommodate a carnivorous rather than an herbivorous diet; and their respiratory system changes to rely on lungs rather than gills.

Studies have shown that the tadpole thyroid gland plays a significant role in inducing these changes. As early as 1912, Gudernatsch demonstrated that anuran tadpoles fed minced thyroid glands would metamorphose prematurely. Normally, a gradual increase in the release of the thyroid hormones thyroxine (T4) and triiodothyronine (T3) stimulates metamorphic change. These hormones apparently act to induce transcription of the genes required for metamorphosis, although the exact mechanism is not known.

Tissues of the tadpole respond differently to the increasing levels of thyroid hormones in a phenomenon referred to as the *threshold effect*. Thyroid hormone concentrations must increase gradually over a long period of time to the threshold concentration for a tadpole to metamorphose normally. If hormone concentrations are abnormally high early in metamorphosis, then normal metamorphosis may be altered. For example, the tail may begin to resorb before the hind limbs have grown in tadpoles exposed to high levels of the thyroid hormones.

Experimental Rationale

In this exercise, you will experimentally manipulate the concentrations of thyroxine and iodine to which tadpoles are exposed and study the effect of actinomycin D, a drug that inhibits RNA synthesis, on tadpole metamorphosis.

II. EXPERIMENTAL PROTOCOL

CAUTION: The drugs used in this exercise are effective on humans. Avoid exposure, and wash hands well after changing solutions.

Materials

Xenopus laevis tadpoles with hind limbs of 1 mm length (at least 40)
Eight dishes or holding tanks
Solutions of thyroxine, iodine, and actinomycin D in well water or 20% Steinberg's solution
Eight dip nets
Dissecting microscope
Metric ruler
Tadpole care and feeding schedule
15-ml screw-cap tubes
5% formaldehyde for fixation

Methods

Your experimental animal for this lab, X. laevis, requires approximately 4 weeks at 22–25°C to achieve the hind-limb stage prior to metamorphosis. For this reason, you may need to start this experiment early in the semester, during Laboratory Exercise 2, Early Amphibian Development. Tadpole care and data collection will continue either until metamorphosis is complete in the control tadpoles or until the end of the semester.

A. Overall Experimental Plan

1. The embryos grow together for approximately 4 weeks until the hind limbs are 1 mm long. Data collection and culture maintenance are performed by all groups during this time.

2. When the hind limbs are 1 mm long, distribute equal numbers of tadpoles into a control group and seven experimental groups (see Table 5.1).

TABLE 5.1

Groups and Media for Tadpole Metamorphosis

Group	Culture solution	Milliliters of stock solution	Milliliters of water
1	Well water	0	100
2	10^{-5} M thyroxine	10	90
3	10^{-6} M thyroxine	1.0	99
4	10^{-7} M thyroxine	0.1	99.9
5	10^{-7} M thyroxine, 1 week[a]	0.1	99.9
6	2×10^{-6} M iodine	2.0	98
7	1×10^{-6} M iodine	1.0	99
8	2.5 μg/ml actinomycin D	1.0	99

[a]Tadpoles in this group are exposed to thyroxine for 1 week and then maintained in water without addition for the remainder of the experiment.

3. Record the number of tadpoles in each group.

4. Clearly label each group of tadpoles with the appropriate solution and date.

5. There should be no more than 25 tadpoles per 500 ml of solution.

6. The solution need not be deep, and the surface of the container should not be obstructed.

7. From this time to the end of the experiment, each lab group is responsible for culture maintenance and data collection of one or more tadpole groups. Data will be posted in the lab room during this time for recording by all class members.

B. Embryo and Tadpole Care and Feeding

1. Keep the embryos and tadpoles at room temperature in well water, dechlorinated tap water, or 20% Steinberg's solution during the entire course of the experiment.

2. During the period from fertilization to hind-limb stage, the water must be changed three times weekly. Be sure not to injure the embryos during this process.

3. When the tadpoles start feeding, they will eat canned or frozen parboiled spinach or collard greens. Be sure not to overfeed them. If food is left over to decay or fouls the tank after feeding, they are being overfed. The water must be changed daily after tadpoles start feeding.

4. After division of the tadpoles into groups, each student group must change the experimental solution daily and continue to feed the tadpoles daily. Be sure to use the correct solution for your experimental group, and do not cross-contaminate the dip nets with the wrong solution.

5. Remove dead tadpoles from the culture and preserve them in a 5% formaldehyde solution. Place each dead tadpole in a 15-ml tube labeled with the date and its experimental group number (see II.C.1).

6. A schedule for culture maintenance and data collection for (a) the time period before division into experimental groups and (b) the time period after distribution of tadpoles into groups will be prepared and posted in the lab room (see II.C.2).

C. Tadpole Observations

From Fertilization to 1-mm Hind-Limb Stage

1. Keep a daily record of (a) time and date; (b) length, including total length from snout to tip of tail, body length from snout to base of tail, and hind-limb length; (c) stage (see Supplementary Readings); (d) morphology; and (e) mortality for the tadpole group.

2. Coordinate data collection with the other student groups in order to cover every weekday until hind-limb bud stage.

3. Use a dissecting microscope and millimeter ruler to measure embryo lengths described above. Measure five embryos and take the average length. Limit the time the embryos spend out of their solutions.

From 1-mm Hind Limb to Complete Metamorphosis

4. Keep a daily record of (a) time and date; (b) length, including total length from snout to tip of tail, body length from snout to base of tail, and hind-limb length; (c) stage (see Supplementary Readings); (d) morphology; and (e) mortality for the tadpole group.

5. Coordinate data collection with the other student groups in order to cover every weekday until metamorphosis is complete.

6. When working with different experimental solutions, take care not to expose tadpoles to an improper solution by means of contaminated glassware or nets.

7. Use a dissecting microscope and millimeter ruler to take measurements of embryo length. Measure five embryos and determine the average length. Limit the time the embryos spend out of their solutions.

8. To slow their movement, place tadpoles in a petri dish containing ice-cold well water and return them as quickly as possible to their container after measurement.

9. Fix a specimen from each group in 5% formaldehyde each week during the experimental-treatment phase of the exercise; place the specimen in a labeled 15-ml tube.

III. QUESTIONS

1. Graph embryo length against time after fertilization for all groups (control, thyroxine and iodine concentrations, and actinomycin D).

2. Graph survival against time after fertilization for all groups.

3. Discuss the results (length, mortality, stage, morphology) obtained from the different concentrations and exposures to thyroxine.

4. Discuss the results obtained from the different concentrations of iodine. Why is iodine used in this experiment?

5. Discuss the results of the actinomycin D experiment. Why is actinomycin D used in this experiment?

6. List the solutions in the order of their effectiveness in stimulating metamorphosis, from most effective to least.

7. What are some of the internal morphological changes that accompany the changes in external morphology as the tadpoles metamorphose into frogs?

IV. SUPPLEMENTARY READINGS

Frieden, E. (1981). The dual role of thyroid hormones in vertebrate development and calorigenesis. *In* "Metamorphosis: A Problem in Developmental Biology," 2nd ed. (L. I. Gilbert and E. Frieden, Eds.), pp. 545–564. Plenum, New York.

Gilbert, S. F. (1994). "Developmental Biology," 4th ed., pp. 717–729. Sinauer, Sunderland, MA.

Gudernatsch, J. F. (1912). Feeding experiments on tadpoles. *Arch. Ent. Mech.* **35**, 457–483.

Kollross, J. J. (1961). Mechanisms of amphibian metamorphosis: Hormones. *Am. Zool.* **1**, 107–114.

6 LABORATORY EXERCISE

CELL–CELL INTERACTIONS DURING SPONGE AGGREGATION

I. INTRODUCTION

In developing and differentiating systems, cells that come into contact can either remain in contact or move away from each other. Similar cells forming a clump may eventually form a tissue. In this exercise, sponges provide a simple model system for demonstrating the phenomena of cell–cell aggregation and cell sorting.

Sponges are primitive animals composed of two cell layers, a skeleton of calcium carbonate or silicate spicules, and sometimes collagen-like fibers of spongin. Even though there are several differentiated cell types in this organism, cells can be dissociated and will reaggregate, reorganize, and eventually reform another functional animal. Similar processes involving the cell surface are also involved in tissue adhesiveness and cell sorting in higher animals.

Sponges can be dispersed into individual cells by either mechanical or chemical means or by a combination of the two. Sponges placed in seawater free of Ca^{2+} and Mg^{2+} (CaMgFSW) slowly disperse into individual cells. The CaMgFSW slowly dissolves the proteoglycan material holding the sponge cells together. Reaggregation of the cells requires addition of Ca^{2+} and Mg^{2+} or production of new proteoglycan.

Experimental Rationale

Part II.A demonstrates the phenomenon of cell–cell aggregation and the effects of temperature on this process.

The experiments in Parts II.B and II.C reveal whether sponge cell aggregation is species-specific and examine the nature of the aggregation factor (AF) that mediates cell aggregation.

In Part II.D, the chemical nature of the AF and the AF binding site will be studied by means of a competition assay with sugars suspected of being involved in aggregation. If the sugar molecules bind to sites on the AF molecules(s) or the cell recognition site, then the sugars may inhibit normal aggregation. The sugar that best inhibits aggregation is most like either the AF or its receptor.

II. EXPERIMENTAL PROTOCOL

Materials

Sponges of two different genera, e.g., *Microciona* and *Cliona*
Seawater (SW)
Calcium/magnesium-free seawater (CaMgFSW)
Two multiwell plates (12 wells each)
4°C incubator
Cheesecloth
Solutions of glucose, galactose, glucosamine, and glucuronic acid
Aggregation factor
Pasteur pipettes and bulbs
Two rotary shakers
Dissecting microscopes
Compound microscopes
Hemacytometers

Methods

CAUTION: Some individuals are allergic to sponges. Take care either to use tongs or, if you use your fingers to disperse the cells, to wear gloves.

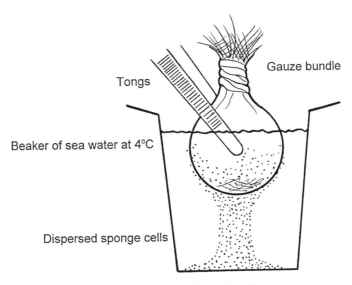

FIGURE 6.1 Dispersion of sponge cells. Cubes of sponge tissue are wrapped in gauze. The gauze bundle is submerged in a beaker of cold seawater and squeezed with tongs (as shown) or with fingers. Examine the dispersed cells by compound microscopy to estimate the degree of dispersion at the start of the experiment.

To disperse sponge cells for use in lab, wrap chunks of sponge tissue in SW in a three-layer-thick square of cheesecloth and massage the "bundle" to release cells into a beaker of cold SW on ice (see Fig. 6.1).

Examine the resulting cell suspension and determine the ratio of cell clumps to single cells in several microscope fields. Also note the size distribution of the newly dissociated cell clumps for comparison to those formed during reaggregation (Fig. 6.2). Use the templates provided in Fig. 6.3 to plan and record your data collection.

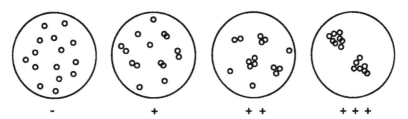

FIGURE 6.2 Degrees of aggregation. Compound microscope field of view for estimation of the formation of aggregates from dispersed sponge cells.

Room temperature incubation

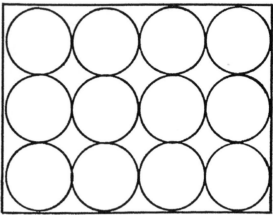

4°C incubation

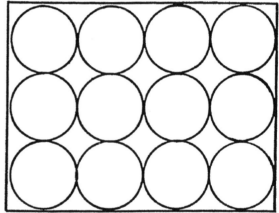

FIGURE 6.3 Twelve-well multiwell plate template. Use these templates to plan your experiments and to record data.

A. Temperature Effects on Aggregation

1. Dispense an aliquot of red sponge cells *(Microciona)* into one well and an aliquot of yellow sponge cells *(Cliona)* into another well on each of your multiwell plates.

2. Incubate one plate at 4°C and one plate at room temperature.

3. Observe the cultures and estimate reaggregation at 0, 15, 30, 60, and 120 min and at 24 and 48 h after mixing.

B. Species-Specific Sorting

1. Mix equal aliquots of the two cell suspensions in a third well of each multiwell plate.
2. Determine by observation whether mixed cell aggregates form in the wells containing cells from both sponges at the two different temperatures.

C. Aggregation Factor and Cell Sorting

1. Preparations of AF have been made from dispersed cells of both *Microciona* and *Cliona*. Add AF from each sponge type to aliquots of each cell type and to a mixture of the two cell types in different wells of your multiwell plate.
2. Incubate at 4°C.
3. Observe whether cell aggregates form in the presence of the AF of the same or different species and record your results.

D. Chemical Nature of AF or the AF Binding Site

1. Place an aliquot of one sponge cell type in each of four wells of your multiwell plate to be incubated at room temperature. Place an aliquot of the other cell type in each of four other wells.
2. To each well of each cell type, add an equal number of drops of one of the four sugar solutions (glucose, galactose, glucuronic acid, and galacturonic acid).
3. Observe whether cell aggregates form in the presence of the different sugars, and record your results.

III. QUESTIONS

The Role of Ca²⁺

1. You have performed several lab exercises using CaMgF media. What is the role of Ca^{2+} in the sea urchin exercise?
2. In the frog exercise?
3. In this exercise?

A. Temperature Effects on Aggregation

1. Why is it important to examine the ratio of clumps to cells in the initial cell suspension?
2. Is there a difference between the times required for cell aggregation at the different temperatures?
3. Why is aggregation observed at room temperature but not at 4°C in the absence of AF?

B. Species-Specific Sorting

1. What would be a reason for the behavior of the cells in the mixed well?
2. Do the aggregates look different in the mixed samples at the two temperatures? Why?

C. Aggregation Factor and Cell Sorting

1. Is the effect of AF species specific? Why?

D. Chemical Nature of the AF or the AF Binding Site

1. What is the chemical nature of sponge aggregation factor and the cell surface receptor?

IV. SUPPLEMENTARY READINGS

Gilbert, S. F. (1994). "Developmental Biology," 4th ed., pp. 28–29. Sinauer, Sunderland, MA.

Humphreys, T. (1963). Chemical dissociation and *in vitro* reconstruction of sponge cell adhesions: I. Isolation and functional demonstration of the components involved. *Dev. Biol.* 8, 27–49.

Wilson, H. V. (1907). On some phenomena of coalescence and regeneration in sponges. *J. Exp. Zool.* 5, 245–258.

APPENDIX A

Laboratory Requirements

MATERIALS NEEDED BY EACH STUDENT

You should obtain the following materials:

Fine forceps
Scalpel
Fine scissors
Dissecting needle
Notebook

KEEPING A LAB NOTEBOOK

You must keep a notebook for all experiments performed in the lab. Bring it with you for each laboratory exercise and whenever you come into lab to collect data.

Because your notebook is the source of all data used in your lab reports, you must keep accurate records. Some hints for keeping a good notebook include:

(a) Start each laboratory exercise on a new page, complete with date and title.

(b) After reading the instructions for the exercise, determine what data will be collected and construct appropriate tables for their collection before coming to the laboratory.

(c) Write down all information and data you collect during the laboratory period in an understandable, logical order. Include time(s), date(s), measurements (with units), sketches, calculations, etc. Enter all information into your notebook directly, not on paper toweling or paper scraps first, for copying into your notebook.

(d) Actually *use* your notebook.

You may not be required to turn in your notebook, but you may be asked to present it to justify statements in your lab reports.

WRITING LAB REPORTS

Your laboratory reports should follow the general guidelines for a scientific paper and include the following sections: Title, Introduction, Results, Conclusions, Projections. Your name, lab section, and date of submission should appear at the beginning of each report. Note that a "Materials and Methods" section, usually included in a scientific paper, is not necessary for your report. A reference to the appropriate section of this manual will substitute for it in your report.

The sections of your report should include the following information:

(a) Title (1 sentence). The title of the laboratory exercise.

(b) Introduction (1–2 paragraphs). The precise question(s) you are asking in the experiment(s). Include the reasoning behind the experimental design, the theoretical results expected, and the experimental controls needed. You need not restate the materials and methods used in the laboratories.

(c) Results (2–3 pages, as needed). Your data and observations presented in an appropriate, easily understood form (graphs, tables, drawings, etc.) with written explanation. Use a straightedge or a computer to prepare all tables and graphs. Label axes clearly, and include a legend for interpretation of the graph.

(d) Conclusions (1–2 pages). The answer(s) to the question(s) you were asking.

(e) Projections (1–2 paragraphs). Describe the experiments you would perform next, on the basis of the information gained from the completed laboratory exercise.

Use the questions at the end of each lab as a guide in writing your reports. Be sure to include in your report all data required to answer the questions, and discuss the questions in your conclusions. Do not, however, include a "question" section in your report.

APPENDIX B

Care and Use
of Microscopes

Traditionally, developmental biologists have relied on microscopes to gather a large part of their experimental data. These data generally consist of observations made on experimentally manipulated developing systems, such as the sea urchin, chick, and frog. Advances in the technical quality of microscopes have led to better observations and deeper understanding of developmental processes at the cellular level in these systems.

As you will use microscopes extensively for data collection during this course, you must become very familiar with their use. A complete understanding of the workings of the compound and dissecting microscopes used in these laboratory exercises will save you both time and eye strain in the weeks ahead.

CARE AND USE OF MICROSCOPES

Care

Handle the microscopes carefully when bringing them to your table. Use two hands when carrying the microscope. Avoid jolts and vibrations, which may damage the optics and light filament.

1. Use only lens paper to clean the lenses. Kimwipes, facial tissues, and paper towels scratch the lenses.

2. Use no strong force on the knobs or levers.

3. Keep the stage clean and dry.

4. Turn off the light, clean and cover the microscope, and return it to the cabinet when you have finished.

5. Report any trouble to the instructor.

Use (Refer to Fig. B.1)

Compound Microscope

1. Place the 10× objective lens in position on the microscope.

2. Place your slide on the stage.

3. Turn the lamp on to a low illumination. Adjust the light with the rheostat.

4. Looking at the microscope stage from the side, move the objective lens with the coarse focus knob until it is just above the sample on the slide. Looking through the eyepieces, focus with the coarse focus knob, *always upward*—that is, always increasing the distance between the slide and the objective lens. Finish focusing with the fine focus knob.

5. Adjust the ocular lenses to your eyes.

6. Close the iris diaphragm on the condenser until it is just outside the perimeter of the field. Adjust for best resolution.

7. The microscopes are parfocal, so you should need only minor adjustment of the focus when you change objective lenses.

Dissecting Microscope

Dissecting microscopes come in quite a variety but share similar features.

1. Find the light switch and the light direction knob. Move the latter to see its effects.

2. Find the magnification knob.

3. Adjust the oculars.

Always focus upward, increasing the distance between the lenses and the sample.

WORKSHOP 1: DETERMINATION OF FIELD DIAMETERS

Compound Microscope

Knowing the diameters of the viewing fields at the different magnifications possible on your compound microscope will help you determine the approximate sizes of the various organisms you will be studying in this course. These

DISSECTING MICROSCOPE frontal view

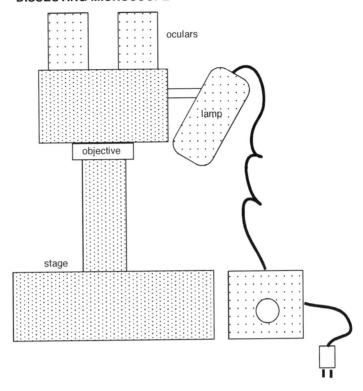

oculars

lamp

objective

stage

COMPOUND MICROSCOPE side view

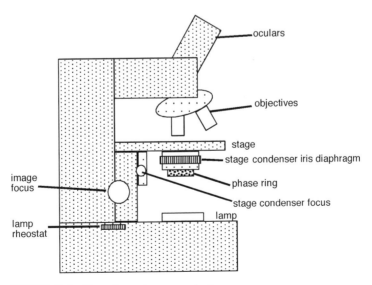

oculars

objectives

stage

stage condenser iris diaphragm

phase ring

image focus

stage condenser focus

lamp

lamp rheostat

FIGURE B.1 Dissecting and compound microscopes.

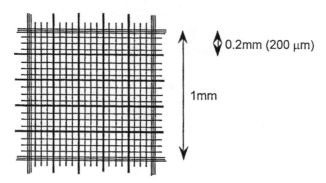

0.2mm (200 μm)

1mm

FIGURE B.2 Center grid of a hemacytometer. The measurements given for this area of a hemacytometer grid can be used to calibrate the compound microscope field at each magnification.

diameters are easily determined with the aid of a hemacytometer according to the following procedure.

1. Turn the objective lens to the lowest magnification and focus the microscope on the reflective portion of the hemacytometer. Etched into the glass of the hemacytometer is a grid of specific dimensions (Fig. B.2).

2. Using the hemacytometer as a ruler, measure the field diameter by aligning one end of the grid along the side of the field. Determine the greatest distance (diameter) across the field to the nearest 0.05 mm (50 μm).

3. Repeat for all magnifications and record the measurements.

Dissecting Microscope

Perform the same type of measurements with the dissecting microscope using a millimeter ruler.

1. Turn the lens to the lowest magnification and focus the microscope on the millimeter ruler placed flat on the stage.

2. Using the ruler, measure the field diameter by aligning one graduation marking along the side of the field. Determine the greatest distance (diameter) across the field to the nearest 0.25 mm (250 μm).

3. Repeat for all magnifications and record the measurements.

WORKSHOP 2: SIZE ESTIMATION

The cells and organisms under study during this course differ widely in their sizes and features. Indeed, some of the cells you will see, such as oocytes, are larger than many entire organisms. Because the units typically used for micro-

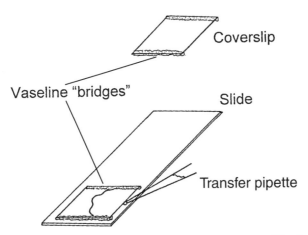

FIGURE B.3 Temporary wet mount preparation. Use of Vaseline "bridges" to elevate the coverslip allows for easy slide preparation and visualization of moderately large cells at high power. Up to three 18- × 18-mm coverslips will fit on a single slide.

scopic measurements (millimeters, micrometers) are probably not part of your everyday life experience, you should become familiar with these units and learn to estimate sizes in terms of them so as better to understand the relationships between cells, the organs they compose, and their relation to the entire organism.

Use the following technique to make a temporary wet mount for sample examination using a compound microscope. Smear a moderate amount of Vaseline on the back of one of your hands, where it will be unlikely to rub off on your papers or slides. Lightly scrape the smear with one edge of a coverslip. Rotate the coverslip 180° and lightly scrape the smear again, to make a coverslip with Vaseline "bridges" as shown in Fig. B.3. Gently place the coverslip, Vaseline side down, on a microscope slide. Surface tension allows liquid samples to be loaded from the side, as shown in Fig. B.3.

You will be provided with a variety of cells and organisms of different sizes and shapes.

1. Place the specimens on slides and estimate their sizes.
2. Use the hemacytometers and rulers to determine the sizes more accurately. Don't forget to include the magnification of the eyepieces in your calculations!

WORKSHOP 3: DETERMINATION OF VOLUMES

It is easy to get "stuck" thinking two dimensionally when using the microscope. When considering differences in cell sizes as in the previous workshop, you probably think first of its diameter (2D) without consideration of the volume difference (3D).

Using the diameters you determined in Workshop 2, determine the approximate volumes of the cells you measured. (Use the relationship $V = 4/3 \times r^3$.)

What is the volume difference between cells differing in diameter by a factor of 10? 100?

APPENDIX C

Gel Electrophoresis

INTRODUCTION

Cells of developing animals express one or several genes in addition to those involved in carrying out the basic "housekeeping" functions of the cells. These genes are often expressed at specific times during development of specific cell types, allowing cells to differentiate into tissues having specialized functions and to express specialized cell products necessary for carrying out those functions. For example, oxygen-carrying blood cells produce hemoglobin, whereas blood cells of the immune system produce antibodies; muscle cells elongate into long fibers and synthesize large amounts of the contractile proteins actin and myosin; neural cells develop cytoplasmic extensions and synthesize neurotransmitters.

If cells express different genes, either at different times or in differentiated states, then some of the proteins in the cells should be different. Comparison of the proteins in an organism at different developmental stages should reveal some proteins common to all stages and all cell types (the "housekeeping" proteins) and some proteins that differ, the stage-specific or tissue-specific proteins.

The advent of molecular biology has provided tools for investigators to use in probing developmental phenomena at the molecular level—tools not imagined even a short time ago. One powerful technique used to study proteins and nucleic acids is gel electrophoresis. SDS–polyacrylamide gel electrophoresis (SDS–PAGE) is one way to examine the complement of proteins from cells in different developmental stages or tissues. Agarose gel electrophoresis is one way to examine the RNAs and genes (DNA) encoding these proteins in different developmental stages or tissues.

Sorting out and characterizing molecules of interest from the complex mixture found in living cells and tissues is exceedingly difficult. Gel electrophoresis is one fractionation technique used by developmental biologists to study and isolate proteins and nucleic acids of interest in cells. An aqueous mixture of macromolecules from the cell or tissue is placed in a gel matrix in an electric field. Molecules having different net charges and sizes migrate at different rates through the matrix and are thus physically separated from each other.

Under certain conditions, the relative mobility of macromolecules through the gel matrix is proportional to the log of their molecular weights. The distance a molecule under study migrates through a matrix can be compared to the migration distance of known molecules in the same matrix and thus reveal the unknown molecule's molecular weight or size.

The following exercises are designed to acquaint you with the equipment and techniques involved in gel electrophoresis, as you will employ electrophoresis in laboratory exercises during this course. You will run two different gels in this exercise: an SDS–polyacrylamide gel, used to separate a mixture of proteins, and an agarose gel, used to separate a mixture of DNA or RNA molecules. After running and staining the molecules in the gels, you will determine the distance the stained molecules have migrated. By extrapolation from the migration distances of known markers run in both gels, you will determine the molecular weights of the stained molecules.

EXERCISES

CAUTIONS

1. Unpolymerized acrylamide is a neurotoxin. Always wear gloves when handling it.

2. Ethidium bromide is a mutagen. Wear gloves when handling it. Be sure to place any contaminated waste in the EtBr waste can for disposal as hazardous waste.

3. *Do not* look at the UV lamp without safety goggles or a face shield. Limit exposure of your skin to UV light.

4. *Do not* attach or detach the leads from the gel apparatus while the power supply is on.

Materials

Protein samples
Molecular weight markers for protein gel

20-μl (microliter) pipettor
Pipette tips
Boiling water bath
Sample support for boiling water bath
SDS–PAGE gel (precast)
Vertical gel apparatus
Electrode or running buffer
Power supply
Large dish for staining and destaining gels
Coomassie blue stain
Destaining solution
DNA samples
Molten agarose
Gel casting tray and comb
Horizontal gel apparatus
Power supply
Electrode buffer
Hand-held UV light source
UV-protective eyewear (goggles or face shield)

A. SDS–Polyacrylamide Gel Electrophoresis (SDS–PAGE) of Protein Samples

Because of time limitations, gels prepared in advance are available for your use in this exercise (see Fig. C.1 and Appendix D). Protein samples to run on the gels are partially prepared, but they must be boiled before use.

1. Prepare a boiling water bath.

2. Carefully float the protein samples in the boiling water bath for 5 min.

3. During this time, mount the precast gel sandwich on the vertical gel apparatus and fill the buffer reservoirs with Tris-glycine running buffer (see Fig. C.2).

4. Load 10 μl of each protein sample into a lane on the gel (see Fig. C.2). Reserve one lane on each gel to load with molecular-weight markers, a mixture of proteins of known molecular weights. Record which sample is loaded in each lane.

5. Carefully place the lid on the apparatus and plug the apparatus into the power supply.

6. Run at 40 mA (20 mA per gel for two gels) until after the dye in the samples reaches the very bottom of the gel (approximately 1 h) (Fig. C.2).

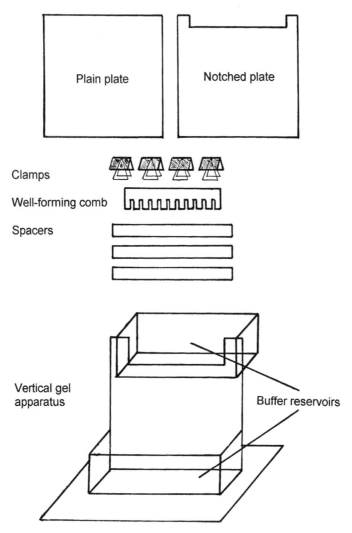

FIGURE C.1 Equipment for SDS–PAGE. Equipment for SDS–PA gel electrophoresis includes two glass plates, butterfly clamps, a well-forming comb and spacers that are 0.5 to 1.5 mm thick for the sides and bottom of the glass plates, and a Plexiglas vertical gel apparatus. This equipment can be purchased commercially or constructed to the desired dimensions.

7. Turn off the power and unplug the apparatus.

8. Remove the plates from the apparatus and gently separate them.

9. Place the gel in a large staining dish containing Coomassie blue protein stain.

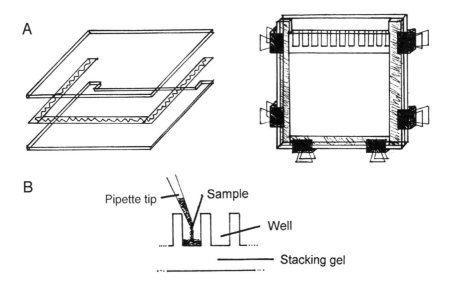

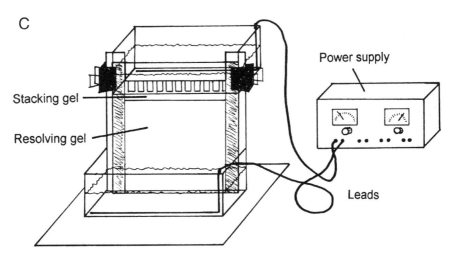

FIGURE C.2 Assembly and electrophoresis of an SDS–PA gel. A. Spacers are coated on both sides with vacuum grease and placed between the notch and plain glass plate as shown. Butterfly clamps hold the plates together while the well-forming comb is positioned at the top of the notched plate. The resolving and stacking gels are poured and polymerized sequentially. B. After the stacking gel has polymerized, the bottom spacer is removed, and the gel–plate "sandwich" is clamped to the vertical gel apparatus. The buffer reservoirs are filled with Tris-glycine running buffer. The well-forming comb is carefully removed, and samples are loaded. C. The gel apparatus is connected to the power supply and the gel is run until the sample buffer dye reaches the bottom of the gel.

10. Stain the gel for 15–30 min or longer while gently shaking the dish on a rotary shaker. Staining time increases with increasing gel thickness and percentage acrylamide.

11. Pour off the stain and rinse the gel several times in destaining solution.

12. Allow the gel to soak in destaining solution until you can see bands in the gel. After approximately 30 min, you may see bands. Full destaining may take several hours.

13. After complete destaining, the gel may be dried on paper or on plastic film for preservation.

Measure the distance of migration for the known molecular-weight markers. Produce a standard curve by plotting the log of the molecular weights of the known markers against the distance migrated through the gel on the three-cycle semilog paper provided at the end of this chapter (see Fig. C.5). Measure the distance of migration of unknown bands in your samples, and determine the molecular weights of the unknowns by extrapolation from the curve of known molecular-weight markers.

B. Agarose Gel Electrophoresis of DNA Samples

Prepare the gel casting tray for pouring an agarose gel (see Fig. C.3); 1× TAE electrophoresis buffer and a mixture of 1% agarose in 1× TAE buffer have been prepared for your use. Melt the agarose by heating it in a microwave oven for a short period of time without allowing the agarose to boil over. After melting, keep the agarose molten in a 65°C water bath until use. The DNA samples for analysis are ready to load onto the gel.

1. Add ethidium bromide to the molten agarose.

2. Pour the agarose into a casting tray with the comb in place and allow the gel to cool and harden undisturbed (see Fig. C.4).

3. After the agarose has hardened, remove the comb and transfer the gel to the electrophoresis chamber.

4. Fill the chamber with electrophoresis buffer until the gel is just covered and the wells are full.

5. Carefully load 10 μl of a sample into each well. Reserve one lane on each gel to load with molecular-weight markers, a mixture of DNA fragments of known sizes. Do not move the gel or the apparatus once you start to load the samples.

6. Place the lid on the chamber and plug the lead wires into the power supply.

7. Run the gels at a constant 100 V for 1 h or until the dye is close to the bottom of the gel.

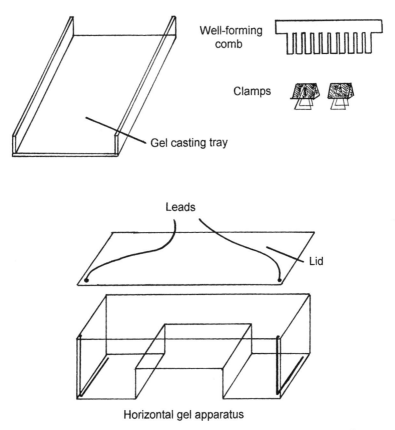

Well-forming comb

Clamps

Gel casting tray

Leads

Lid

Horizontal gel apparatus

FIGURE C.3 Equipment for agarose gel electrophoresis. Equipment for agarose gel electrophoresis includes a gel casting tray that is open on two ends, a well-forming comb, butterfly clamps, and a horizontal gel apparatus. Equipment can be purchased commercially or constructed to the desired dimensions.

Visualize the DNA bands in the gel by illumination with UV light. Measure the distances of migration for the known and unknown bands. Produce a standard curve by plotting the length (in base pairs) of the known DNA fragments against the distance migrated on semilog paper (see Fig. C.5 for a sample of semilog paper and Appendix D for the sizes of molecular-weight markers). Determine the lengths of the unknowns by extrapolation from the standard curve.

A

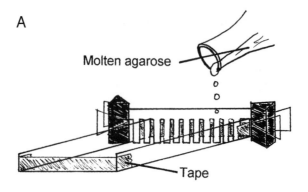

Molten agarose

Tape

B

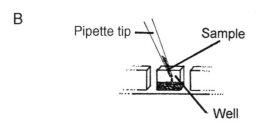

Pipette tip

Sample

Well

C

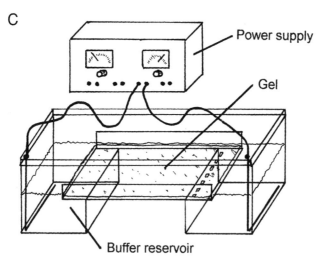

Power supply

Gel

Buffer reservoir

FIGURE C.4 Assembly and electrophoresis of an agarose gel. A. The open ends of the gel casting tray are sealed with tape (scotch tape works). The well-forming comb is positioned so that the bottoms of the teeth are about 1–2 mm above the surface of the gel casting tray. Molten agarose cooled to <65°C is then poured into the casting tray. B. After the gel solidifies, the tape is removed from the ends of the tray, the gel still on its tray is positioned in the horizontal gel apparatus, and the samples are loaded. C. After the gel is loaded, the lid and leads are connected to the power supply, and the gel is run for about 1 h.

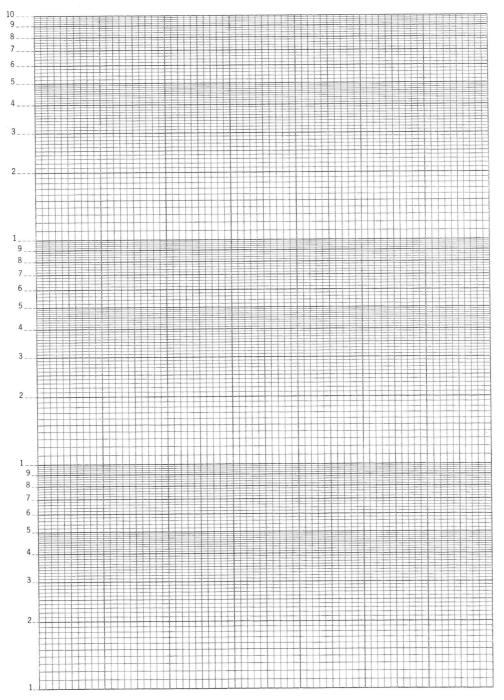

FIGURE C.5 Three-cycle semilog paper for size determination. Using the semilog graph paper, plot distance migrated against molecular weight of known molecules. Draw a line to connect the points on the curve. Use the curve to extrapolate the molecular weights of unknown samples.

WORKSHOP 1: PLASMID MAPPING

Sample Problem: Restriction Mapping of a DNA Plasmid

Plasmids are closed circles of extrachromosomal DNA that are used by investigators for a variety of molecular-biology techniques.

Restriction mapping uses restriction endonucleases to cleave DNA at specific sites, thereby producing DNA fragments of varying lengths. By determining the length(s) of the different DNA fragment(s) produced by different restriction enzymes alone and in combination, it is possible to map their cleavage sites on the plasmid. One may need to know this information to construct clones, sequence DNA, etc., using the plasmid.

Instructions

Your task is to construct, from the data given in Table C.1, a restriction map of a theoretical plasmid with the restriction sites and the distances between the sites labeled. To do so, you will first graph data on a set of DNA fragments of known lengths. From these points you will be able to produce a standard curve, the slope of which represents DNA fragment length/distance migrated on the gel.

TABLE C.1

Data from Which to Construct a Hypothetical
Restriction Map

Knowns	
DNA length (base pairs)	Migration distance (mm)
20,000	10
5,000	40
2,000	60
800	80
500	90
Restriction digests	
Enzyme(s) used	Migration distance (mm)
A alone	36
B alone	44 and 60
C alone	36
A and B together	60
A and C together	51
A, B, and C together	60 and 75

Using the standard curve, you will be able to determine the lengths of DNA fragments produced by a series of restriction-enzyme (A, B, and C) digests of a fictional plasmid, given their migration distances in a gel. From the lengths of the different products of digestion, you will produce the plasmid's restriction map.

To produce the graph of DNA lengths versus distance migrated, you will need a sheet of three-cycle semilogarithmic graph paper (see Fig. C.5).

APPENDIX D

Guide to Preparation for Laboratory Exercises

I. GENERAL LIST OF EQUIPMENT AND SUPPLIES FOR LABORATORY EXERCISES

Each item is followed by a listing, in parentheses, of the laboratory exercises in which it is used.

A. Equipment

Dissecting microscopes (1, 2, 3, 4, 6, Appendix B)
Compound microscopes (1, 2, 4, 6, Appendix B)
Clinical centrifuges (1)
Hot plate (3, Appendix C)
pH meter (preparation only)
Test-tube racks (1, 2, 4, 6, Appendix C)
SDS–polyacrylamide gel electrophoresis equipment (1, 4, Appendix C)
Agarose gel electrophoresis equipment (Appendix C)
Power supply for electrophoresis (1, 4, Appendix C)
Hand-held UV light source (2, Appendix C)
UV-protective goggles or face shield (2, Appendix C)
Microliter pipettor (1, 5, Appendix C)
Operational sea tanks (two) (1, 6)
Incubator to operate between 4 and 37°C (1, 3, 4, 6)
Humidified chick egg incubator (3)
High-speed centrifuge and rotor (6)
Bunsen burners and strikers (2, 3, 4)

B. Standard Chemicals

NH_4Cl (ammonium chloride)
$CaCl_2$–2-H_2 (calcium chloride)
$Ca(NO_3)_2$–4-H_2O (calcium nitrate)
Ethanol
EDTA (ethylenediaminetetraacetic acid)
EGTA (ethylene glycol-bis(β-aminoethyl ether) $N,N,N',_2N'$-tetraacetic acid)
Glacial acetic acid
Glycerol
HCl (hydrochloric acid)
Hepes (-N-[2-hydroxyethyl]piperazine-N'-[2-ethanesulfonic acid])
$MgCl_2$–6-H_2O (magnesium chloride)
$MgSO_4$–7-H_2O (magnesium sulfate)
Methanol
$KHCO_3$ (potassium bicarbonate)
KCl (potassium chloride)
$NaO_2H_3C_2$ (sodium acetate)
$NaHCO_3$ (sodium bicarbonate)
NaCl (sodium chloride)
Tris base (tris(hydroxymethyl)aminomethane)

C. Special Chemicals

A23187, Ca^{2+} ionophore (1)
Acrylamide (1, 4, Appendix C)
Actinomycin D (5)
Agar (3)
Agarose (Appendix C)
Agarose gel molecular-weight markers (Appendix C)
Bis acrylamide (1, 4, Appendix C)
Bromophenol blue dye or phenol red dye (1, 4, Appendix C)
Coomassie brilliant blue R dye (1, 4, Appendix C)
Dimethylsulfoxide (DMSO) (1)
Drosophila food (4)
Ether (4)
Ethidium bromide (Appendix C)
Formaldehyde (5)
Galactose (6)
Galacturonic acid (6)
Glucosamine (6)
Glucose (3, 6)

Glucuronic acid (6)
Glycine (1, 4, Appendix C)
Human chorionic gonadotropin (2)
Instant Ocean™ (optional) (1, 6)
Iodine (5)
Orcein dye (4)
Penicillin (1, 3, 6)
Plasmid DNA (Appendix C)
Two restriction endonucleases (Appendix C)
SDS–PA gel molecular-weight markers (1, 4, Appendix C)
Streptomycin (1, 3, 6)
Tadpole food (see Laboratory Exercise 5) (5)
Thyroxine (5)
Trypsin (1)
Trypsin inhibitor (1)
Trypan blue (3)

D. Special Supplies

Twelve-well multiwell plates (1, 6)
Small files (3)
Millimeter ruler fragments (1, 2, 3, 4, 5, Appendix B)
Bottles for culturing *Drosophila* (4)
Gauze (6)
Cotton (3)
No. 1 filter paper (1, 2, 3, 4, Appendix C)
Fine watercolor paintbrushes (4)
35-mm petri dishes (2, 3,)
60-mm petri dishes (2, 3)
100-mm petri dishes (2, 3)
15-ml disposable plastic conical tubes (1, 5)
1.5-ml plastic microfuge tubes with attached lids (1, 4, Appendix C)
Pasteur pipettes (all labs)
Bulbs (all labs)
Plastic tips for micropipettors (1, 4, Appendix C)
Vaseline (all labs)
Microscope slides (1, 2, 4, 6, Appendix B)
Coverslips (1, 2, 4, 6, Appendix B)
Duco cement (3)
Lens paper (all labs)
Wax (3)
India ink (1)

Waterproof marking pens (all labs)
Parafilm (3, preparation)
Foil (preparation only)
Labeling tape (all labs)
Scotch tape (all labs)
Wide package-sealing tape (3)
Gloves (some students may need these)
Squirt bottles (all labs)
Glass capillary tubes (2, 3, 4)
Prepared slides of chick and frog embryos (2, 3)
Test-tube brushes (all labs)
Paper towels (all labs)
Dipnets (5)
Toothpicks (Appendix B, Appendix C)
Nylon screen (1)
Seeds (bean, radish, mustard, celery) (Appendix B)
Plastic boxes for housing tadpoles (5)
Gel staining/destaining dish (1, 4, Appendix C)
Syringes and fine-gauge needles (1, 2)

E. Live Animals

Sea urchins (two species) (1)
Frogs *(Xenopus laevis)* (3)
Fertile, unincubated chicken eggs (3)
Sponges (two species) (6)
Drosophila (4)
Frog tadpoles (5)
Axolotls (optional) (2)

II. GUIDE TO PREPARATION, LABORATORY EXERCISE 1: SEA URCHIN FERTILIZATION

A. Equipment and Supplies List

1. Equipment for Preparation of Lab

Seawater tanks, one at about 65°F for Gulf coast urchins and another at 50–55°F for Pacific and North Atlantic coast urchins
Incubator set at 15–18°C for growth of Pacific and North Atlantic urchin embryos

Two small air pumps and tubing to aerate cultures of embryos of each species

Clinical centrifuges, one for every 4–6 students

SDS–PA gel electrophoresis equipment (optional, for Part II.G): gel, apparatus, power supply, buffers, boiling water bath, as described in Appendix C)

Nylon screen affixed to bottomless beaker with rubber bands

About 12 250-ml plastic beakers in which to collect eggs

About five 5- or 10-ml syringes and many 20-gauge needles for urchin injection

2. Chemicals and Materials for Preparation of the Lab

Standard: NaCl, KCl, $MgCl_2$–6-H_2O, $MgSO_4$–7-H_2O, $CaCl_2$–2-H_2O, $NaHCO_3$, EGTA, glycerol, $KHCO_3$, NH_4Cl

Special: trypsin inhibitor, trypsin, calcium ionophore A23187 dissolved in DMSO, penicillin and streptomycin antibiotic solutions, india ink

Live animals: two species of sea urchins

3. Equipment and Supplies Needed by Each Student (See Materials Section of Laboratory Exercise 1)

B. Timeline

1. Several Months before the Lab

a. Set up seawater tanks to house the urchins and allow them to equilibrate to the appropriate temperature. Introduce fish or other live sea animals into the tanks to facilitate establishment of conditions for control of nitrogen fluxes after introduction of the urchins.

2. Several Weeks before the Lab

a. Arrange for shipment of two species of gravid sea urchins to be delivered as close as possible to the date of the lab.

3. The Day before the Lab

a. Prepare seawaters and other solutions. Prepare artificial seawater (SW) from fresh synthetic sea salts, such as Instant Ocean, or prepare according to Table D.1. Adjust pH to 8.0–8.2 and specific gravity to 1.020–1.023 at 75°C. Do not use SW from the urchin holding tanks, in case any urchins have

TABLE D.1

Preparation of Artificial Sea Waters

Chemical	Formula weight (g/mol)	g/1 Liter	g/2 Liters	Final molarity (mM)
A. Artificial sea water (SW)				
NaCl	58.44	24.72	49.44	423.0
KCl	74.5	0.67	1.34	9.0
$MgCl_2-6-H_2O$	203.3	4.68	9.35	23.0
$MgSO_4-7-H_2O$	246.5	6.4	12.8	26.0
$CaCl_2-2-H_2O$	147.0	21.36	2.73	9.27
Dissolve the above salts in deionized H_2O, then add:				
$NaHCO_3$	84.01	0.185	0.374	2.2
Adjust the pH to 8.0–8.2, bring to volume with H_2O, and add antibiotics as described above.				
B. Calcium/magnesium-free sea water (CaMgFSW)				
NaCL	58.44	30.7	61.5	526
KCl	74.5	0.745	1.49	10
EGTA	380.35	0.76	1.52	2
$NaHCO_3$	84.01	0.21	0.42	2.5
Dissolve salts in deionized H_2O, bring to volume with H_2O, and add antibiotics as described above.				
C. Sodium-free sea water (NaFSW)				
Glycerol	92.1	84.7	169.5	920
KCl	74.5	0.745	1.49	10
$CaCl_2-2-H_2O$	147.02	1.62	3.23	11
$MgCl_2-6-H_2O$	203.3	5.29	10.57	26
$MgSO_4-7-H_2O$	246.5	7.15	14.3	29
$KHCO_3$	100.12	0.20	0.04	2
D. 0.55 M KCl solution				
KCl	74.5	41.0		550
E. 0.65 NH_4Cl solution				
NH_4Cl	53.49	34.8		650

shed their gametes into the water. Add 10^5 U/liter penicillin and 100 mg/liter streptomycin to increase viability of the embryos.

b. For each lab bench, label a set of tubes for sperm and special samples to be provided during the lab. Do not underestimate the time required for this lengthy task. Color code the tubes for each species and solution. Once labeled, tubes can be washed with the labels in place for use the next time the lab is taught. For example, label the tubes as follows:

Species 1 sperm in SW
Species 2 sperm in SW

Species 1 sperm in CaMgFSW
Species 2 sperm in CaMgFSW
Species 1 sperm in NaFSW
Species 2 sperm in NaFSW
Species 1 jelly water
Species 2 jelly water
Protease in SW
Protease inhibitor in SW
Ionophore in SW
Ionophore in NaFSW
Ionophore in CaMgFSW

c. For each lab bench, label a set of beakers for eggs:

Species 1 eggs, washed
Species 2 eggs, washed
Species 1 eggs, unwashed
Species 2 eggs, unwashed

d. Set out equipment and supplies at each lab bench:

1.5-ml tubes of india ink, diluted with 1 vol H_2O
Slides and coverslips
Vaseline
Glass Pasteur pipettes and rubber bulbs, or plastic transfer pipettes
Two 12-well plastic multiwell culture plates per student
Waterproof marking pens
Kimwipes
Hemacytometers
100-ml bottles of SW, CaMgFSW, NaFSW, and NH_4Cl
15-ml conical plastic centrifuge tubes, 3–5 per student
Racks for 15-ml centrifuge tubes
A clinical centrifuge for 15-ml tubes at each bench
Pans with very dilute soapy water for student clean-up

e. Prepare "gamete collecting kits," each on a tray or in a pan:

1–2 petri dishes for males
6–8 250-ml plastic beakers for females
3–4 10-ml luerlock syringes
Many 20-g luerlock needles
Several liters of SW (do not use SW from the tanks)
Paper towels
Plastic bag for urchin carcasses

4. The Day of the Lab

a. Distribute the seawaters and solutions into the appropriate tubes and beakers of each set (steps 3.c and 3.d).

b. At each bench, set out ice buckets for gametes and pans of very dilute soapy water and brushes for washing dishes.

c. Be prepared to make the following solutions as needed by students:

Protease inhibitor (crude soybean trypsin inhibitor): 0.4 mg/ml in SW
= 20 mg/50 ml SW
= 6 mg/15 ml SW
Distribute into 15-ml tubes, and place one tube on ice at each lab bench.
Protease (trypsin): 0.5% in SW
= 0.5 g/100 ml SW
= 0.25 g/50 ml SW
Distribute into 15-ml tubes, and place one tube on ice at each lab bench.
Ionophore A23 187
Dissolve the ionophore in DMSO to make a 2-mg/ml stock solution that is
    stable for several months when stored at 4°C. Prepare working solutions
    by diluting 0.15 ml of the stock in 15 ml of each kind of SW.

d. Collect gametes from sea urchins as follows:

(1) Insert a syringe needle through the soft tissue around the mouth on the ventral side of the urchin, and inject about 5 ml of 0.55 M KCl into the coelomic cavity of the urchin (see Fig. D.1C). Gently shake the urchin to distribute the KCl, and place it dorsal-side-up on a clean paper towel. Rinse the needle and syringe thoroughly or use a fresh needle and syringe for injection of each urchin to avoid exposing females to concentrated sperm samples.

(2) Determine the sex of the urchin by observing the color of the gametes exuded from the gonopores on the dorsal side of the urchin. A white milky exudate is generally produced by males of all species; *Strongylocentrotus purpuratus* females exude pale orange eggs; *Arbacia punctulata* females exude purple eggs; *Lytechinus pictus* eggs are pale yellow. If in doubt, examine a sample of the exudate using a compound microscope.

(3) Place the males on a petri dish or paper towel and continue injecting individual urchins until the desired number of females has been obtained.

(4) Fill a 250-ml beaker with fresh SW, place a shedding female dorsal-side-down on the beaker, and allow the eggs to drift to the bottom of the beaker (see Fig. D.2).

(5) For each lab bench, set aside about 15 ml of egg suspension of each species without washing them further, for use as "eggs with jelly" in

A

Dorsal side

Spines

Shell

Ventral side

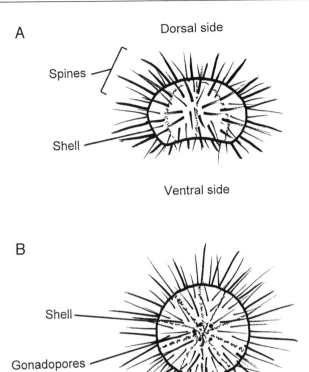

B

Shell

Gonadopores

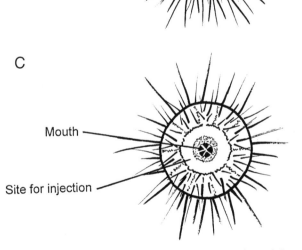

C

Mouth

Site for injection

FIGURE D.1 Sea urchin anatomy. A. Side view of an adult sea urchin, showing the shell, or test, and spines. B. View from the dorsal side of a sea urchin, indicating the gonopores through which gametes are shed. C. View from the ventral side of a sea urchin showing the mouth and indicating the site for injection of KCl solution to induce shedding of gametes.

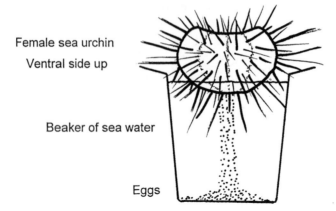

Female sea urchin
Ventral side up

Beaker of sea water

Eggs

FIGURE D.2 Sea urchin egg collection. After injection with 0.55 M KCl, the female urchin is placed with its ventral side up over a beaker of seawater, so that eggs released through the gonopores on the dorsal side are collected in fresh seawater.

Part II.A of the lab. For parts II.B through II.F of the lab, dejelly the remainder of the eggs by passing them through a fine mesh (100 ct) nylon screen several times or by repeated settling and resuspension in CaMgFSW. Dejellied eggs are usually more synchronous in their responses than eggs that retain their jelly coats. Maintain the beakers of Gulf coast eggs at room temperature, swirling them occasionally to aerate the suspension. Maintain beakers of Pacific and North Atlantic eggs in a sea tank, water bath, or incubator at about 15–18°C.

(6) Save SW from the first wash as "jelly water" for Part II.B of the lab.

(7) To obtain sperm, cut open the carcasses with scissors, dissect the testes from the coelom of one or two males of each species, and store them "dry" in a petri dish at 4°C with as little SW or fluid as possible. Sperm stored "dry" at 4°C in this manner are usable for up to a week after dissection.

(8) At the beginning of the lab, suspend about 10 drops of "dry" sperm in 10 ml of the appropriate SW to make a concentrated sperm suspension, and store on ice. Dilute 10 drops of concentrated sperm suspension into 10 ml of the appropriate SW and distribute to students. Remake fresh sperm suspensions from the concentrated sperm suspension every 20–30 min during lab as needed.

e. If the optional sea urchin gel-electrophoresis lab will be performed, note the time and fertilize a batch of eggs of each species, incubate cultures at the appropriate temperature with aeration, and collect the following samples:

Sample approximate collection time
1. Sperm before fertilization
2. Unfertilized eggs before fertilization
3. 1 min after fertilization
4. 2-cell stage 90 min after fertilization
5. morula 3–5 h after fertilization
6. ciliated blastula 12 h after fertilization
7. gastrula 24 h after fertilization
8. pluteus larvae 48 h after fertilization

At collection, pipet 3–5 ml of culture into a centrifuge tube, pellet the gametes or embryos by a low-speed spin, decant the SW, resuspend the pellet in about 1–5 vol of 2× SDS–PAGE sample buffer, boil for 3 min, and freeze until all samples have been collected and the gel is ready.

C. References

Costello, D. P., Davidson, M. E., Eggers, S., Fox, M. H., and Henley, C. (1957). "Methods for Obtaining and Handling Marine Eggs and Embryos." Marine Biological Laboratory, Woods Hole, MA.

D. Resources

1. Sea Urchins

Any of several species of urchins can be used, depending on the time the lab is offered. For a course taught in the spring, use of *S. purpuratus* and *A. punctulata* (ripe from December through April) is recommended. For a course taught in the summer or fall, use of *L. pictus* is recommended. Call the supplier to determine availability of ripe urchins at the desired time.

Marinus, Inc.
1500 Pier C Street
Long Beach, CA 90813
Telephone: (310) 435-6522

Pacific Bio-Marine Laboratories, Inc.
P.O. Box 536
Venice, CA 90291
Telephone: (213) 822-5757

Gulf Marine Specimens
P.O. Box 237
Panacea, FL 32346
Telephone: (850) 984-5297

Animal Supply House
Marine Biological Laboratory
Woods Hole, MA 02543
Telephone: (617) 548-3705, ext. 325

2. Nylon Screen

BioDesign Inc.
P.O. Box 1050
Carmel, NY 10512
Telephone: (914) 454-6610
Fax: (914) 454-6077
Cat. No. NSA: Sampler kit containing 200, 100, 50, 25, and 8 mesh nylon screen
Cat. No. N200S: Kit of three squares of 200 μm nylon mesh
Cat. No. N100S: Kit of three squares of 100 μm nylon mesh

3. Supplies for Seawater Tanks

Synthetic sea salts and supplies for seawater tanks can be obtained from scientific supply houses or from local pet stores.

4. Special Chemicals

Trypsin, trypsin inhibitor, A23187, penicillin and streptomycin, DMSO, and EGTA all can be obtained from:

Sigma Chemical Co.
P.O. Box 14508
St. Louis, MO 63178
Telephone: (800) 325-3010

III. GUIDE TO PREPARATION, LABORATORY EXERCISE 2: EARLY AMPHIBIAN DEVELOPMENT

A. Equipment and Supplies List

1. Equipment for Preparation of Lab

25-gauge needles and 1 ml luerlock syringes
Microscope slides

Single-edged razor blades to separate egg "ribbons" and to mince testes
Petri dishes (100- or 150-mm dishes for egg collection and fertilization)
250- and 500-ml graduated cylinders for preparation of solutions to remove egg jelly

2. Chemicals and Materials for Preparation of the Lab

Standard: $CaCl_2-2-H_2O$, NaCl, KCl, $CaNO_3-4-H_2O$, $MgSO_4-7-H_2O$, Tris base, HCl, NaOH

Special: Human chorionic gonadotropin, dithiothreitol (DTT), Hepes, colored modeling clay, 1.5- × 1.5-cm squares of No. 1 filter paper

Live animals: male and female frogs *(Xenopus laevis)*, axolotl embryos (optional)

3. Equipment and Supplies Needed by Each Student (See Materials Section of Laboratory Exercise 2)

B. Timeline

1. Before the Course Begins

a. Order frogs to ensure availability on the desired date. About 1 week before the lab date, arrange for aquaria or holding tanks to house the frogs upon their arrival. Conditions for housing and handling frogs should be approved by campus Animal Care and Use Committee.

2. Several Days before the Lab

a. Prepare solutions.

(1) 1 mM $CaCl_2$ stock: Dissolve 14.7 mg of $CaCl_2-2H_2O$, in H_2O and adjust volume to 100 ml. (147 g/mol) (0.001 mol/liter) (0.1 liter) = 0.0147 g = 14.7 mg.

(2) Prepare 100% Steinberg's solution according to Table D.2.

b. Purchase modeling clay, or make it according to the following recipe:

3 cups flour
1.5 cups salt
6 teaspoons cream of tartar
3 cups H_2O
3 tablespoons vegetable oil
Red, green, blue, and yellow food coloring

TABLE D.2

Preparation of 100% Steinberg's Solution

Chemical	Formula weight (g/mol)	g/1 Liter	Final molarity (mM) 100%	70%	20%
NaCl	58.44	3.4	58.2	40.7	11.6
KCl	74.5	0.05	0.67	0.47	0.13
$CaNO_3-4-H_2O$	236.2	0.08	0.34	0.24	0.07
$MgSO_4-7-H_2O$	246.5	0.204	0.83	0.58	0.17
Tris	121.1	0.121	1	0.7	0.3

Note. Dissolve the chemicals listed above in 800 ml of H_2O; adjust the pH to 7.4 and the volume to 1 l. According to need, the stock of 100% Steinberg's solution is either used directly or diluted to 70% and 20% concentrations.

Mix dry ingredients together. Mix wet ingredients together. Heat wet ingredients, then slowly add dry ingredients. Stir until well-mixed and a ball of dough forms. Turn out of pan onto a counter. Knead for 3–5 min until smooth. Divide dough into fourths and mix in food coloring. The recipe makes an amount sufficient for about 10 students.

3. The Day before the Lab

a. Prepare to induce ovulation in female frogs. Two female frogs should produce more than enough eggs for 20 students to perform the experiments in lab. To induce ovulation, inject the frogs with human chorionic gonadotropin about 12 to 15 h before the time the eggs are needed for lab, as described below.

(1) Dissolve human chorionic gonadotropin in sterile H_2O at a concentration of 1000 U/ml. Sedate the frogs by incubating them at 4°C.
(2) Inject 1.5 ml of 1000 U/ml human chorionic gonadotropin into the dorsal lymph sacs of each gravid female frog. To ensure that the HGC is delivered to the dorsal lymph sacs, use a fine-gauge syringe needle oriented so that the beveled edge is facing upward, as close underneath the surface of the dorsal skin as possible, on one side of the frog. For injection of *Xenopus* females, the needle should enter the skin approximately 0.5 in. away from the midline of the frog along the "stitch marks" above the leg and should point toward the head. Successful injection will result in temporary appearance of a "bubble" of liquid under the skin. Isolate each injected female in a separate cage filled with

several inches of dechlorinated tap water or well water, and house the injected frogs in a darkened room overnight. Avoid overstimulation of the injected frogs.

4. The Day of Lab

a. Collect sperm from male frogs.

Before collecting eggs from the females, dissect the testes from a male frog, and store them whole "dry" in a covered petri dish at 4°C. Prepare a sperm suspension by macerating one-half of a single testis with fine forceps in a minimal volume of 20% Steinberg's solution. Once the tissue is finely minced, dilute the sperm suspension to 3–5 ml total volume.

b. Collect some blood from the male, and suspend it in 100% Steinberg's solution for use in Part II.F, *Parthenogenesis*.

c. Strip eggs from a female frog.

Approximately 2 h before the beginning of the lab period, collect eggs from each injected female. To collect eggs, hold the female with her dorsal side against the palm of your hand, your forefinger between the legs, and your thumb and middle finger encircling the body just anterior to the hind limbs. Grasp and extend her hind limbs with your other hand. Position the legs in a bent position so that the cloaca is accessible for collection of the eggs. Force eggs from the cloaca by applying gentle pressure to the anterior part of the body, and then progressively working your hand down toward the cloaca. Discard the first few eggs that appear, and wipe the region dry with a Kimwipe. Then continue "milking" the female, depositing the eggs in several "ribbons" onto clean microscope slides. The female should be returned to the holding tank after egg collection. Females will continue to ovulate over the next 6–8 h and can be "remilked" for eggs as needed.

Fertilize enough eggs for students to perform Parts II.A, II.D, II.E, and II.F of the laboratory exercise. Be prepared to collect additional eggs during lab for students to use in performing Parts II.A, II.B, II.C, and II.F. Be ready to prepare fresh sperm suspension using the other half testis for Parts II.A, II.B, II.C, and II.F.

For Parts II.A, II.D, II.E, and II.F, eggs should be collected and fertilized 1.5–2 h before the beginning of the lab period, so that they will be near first cleavage when the students are ready for them during the exercise. To fertilize the eggs, place the microscope slides containing them in petri dishes and drizzle sperm suspension over the freshly deposited eggs. Incubate the eggs in sperm suspension for 15 min, then pour off the sperm suspension and flood the eggs with well water or 20% Steinberg's solution. A successfully fertilized egg will reorient itself with its dark animal cap upward within 30 min of fertilization.

d. Prepare the 8 mM DTT dejellying solution and prepare Steinberg's solutions for washing eggs free of DTT after the jelly is removed.

(1) Dissolve 1.54 g DTT in 10 ml distilled H_2O to make a 1 M stock. Dissolve 23.83 g of Hepes (free acid) in 100 ml of H_2O and adjust the pH to 8.9 with NaOH to make a stock of 1 M Hepes, pH 8.9.
(2) To prepare 8 mM DTT dejellying solution, mix 1.6 ml of 1 M DTT and 12 ml 1 M Hepes, and adjust the volume to 200 ml with distilled H_2O.
(3) 500 ml of 100% Steinberg's solution
(4) 200 ml of 70% Steinberg's solution
(5) 200 ml of 20% Steinberg's solution

e. Set out prepared slides for student use.

C. Resources

1. Special Chemicals

Human chorionic gonadotropin, DTT, and Hepes are available from
Sigma Chemical Co.
P.O. Box 14508
St. Louis, MO 63178-9916
Telephone: (800) 325-3010

2. Frogs

Gravid female and male *Xenopus* frogs can be obtained from the following supply houses:

NASCO
901 Janesville Ave.
Fort Akinson, WI 53538
Telephone: (800) 558-9595

Xenopus I
716 Northside
Ann Arbor, MI 48105
Telephone: (313) 426-2083

Ward's Natural Science Establishment
5100 West Henrietta Road
P.O. Box 92912
Rochester, NY 14692-9012
Telephone: (176) 359-2502

Carolina Biological Supply Company
2700 York Road
Burlington, NC 27215
Telephone: (919) 584-0381

3. Axolotls

Axolotls can be obtained from:

IU Axolotl Colony
Jordan Hall 407
Bloomington, IN 47405
Telephone: (812) 855-8260
Fax: (812) 855-6705
E-mail: sjborlan@indiana.edu

4. Prepared Slides

Prepared slides can be obtained from many biological-materials supply houses, for example:

Carolina Biological Supply Company
2700 York Road
Burlington, NC 27215
Telephone: (919) 584-0381

Ward's Natural Science Establishment
5100 West Henrietta Road
P.O. Box 92912
Rochester, NY 14692-9012
Telephone: (176) 359-2502

IV. GUIDE TO PREPARATION, LABORATORY EXERCISE 3: EMBRYONIC CHICK DEVELOPMENT

A. Equipment and Supplies List

1. Equipment for Preparation of Lab

Temperature- and humidity-controlled egg incubator
12–15°C incubator to hold eggs until beginning of incubation
Hot plates
Small files
Bunsen burner

Organ-culture dishes (see directions for preparation below)
Sterile 13×100 petri dishes
Rulers
Capillary tubes for making fine glass knives
Egg cartons to hold eggs during incubation
250-ml beakers and squirt bottles
Kimwipes

2. Chemicals and Materials for Preparation of the Lab

Standard: NaCl, CaCl2–2H$_2$O, KCl, glucose, agar

Special: Penicillin and streptomycin antibiotics, egg albumen (from extra eggs), No. 1 filter paper, melted wax, 22-×-22-mm coverslips, 70% ethanol, sterile cotton-plugged Pasteur pipettes and bulbs, Parafilm, clear wide packing tape (optional: for sealing windows of eggs)

Live animals: fertile chicken eggs, unincubated

3. Equipment and Supplies Needed by Each Student

See Materials section of Laboratory Exercise 3. Each student will also need fertile chicken eggs incubated the following number of days for the following parts of the exercise:

Unincubated for Part II.B.
Incubated 3–4 days for Part II.C and Part II.D.
Incubated 11–12 days for Part II.D.
Incubated 4–5 days for Part II.E.
Incubated to the 6–8 somite stage, 27–29 h, for Part II.F.

B. Timeline

1. Before the Course Begins

a. Locate a source of fertilized chicken eggs. These can be obtained locally or from the sources listed below.

b. Calculate how many eggs are needed for each part of the lab. A recommendation is to run the lab for 3 weeks and to perform Parts II.A and II.B during week 1; parts II.C, II.D, and II.E during week 2; and part II.F during week 3. The eggs should arrive fresh just before incubation time but can be held for less than a week at 15°C before incubation at 37°C. Order extra eggs, so that students have plenty of material to work with in this exercise. Use Table D.3 to plan the egg order and incubation schedule.

TABLE D.3

Chart for Use in Determining Number of Eggs To Order and Incubation Schedule for Laboratory Exercise 3

	Part of laboratory exercise				
	II.B	II.C	II.D	II.E	II.F
Minimum needed per student:	1	1	2	1	2
Recommended number to order:	3	4	4	2	5
Number of eggs ordered (dozen):					
Time to incubate prior to lab:	1–2 days	4–5 days	11–12 days (2) 4–5 days (2)	4–5 days	27–29 h
Total number to incubate (dozen):					
Date to start incubation: (filled in for each semester)					30–40 h (determine empirically)

2. Several Weeks before the Lab

a. Equilibrate temperature and humidity in the egg incubator.

b. Begin preparation of organ-culture dishes. Cement a 35-mm sterile plastic petri dish bottom into the bottom of a 60-mm plastic petri dish, being especially careful to keep the inner 35-mm dish sterile. Allow plenty of time for the cement vapors to dissipate before the dishes are needed for lab. Make two dishes per student, plus a few extras. Figure D.3 shows a diagram of a completed organ-culture dish.

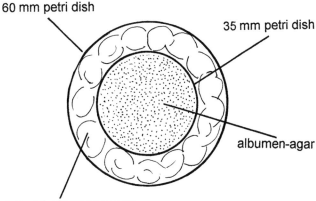

60 mm petri dish

35 mm petri dish

albumen-agar

Moistened cotton or paper

FIGURE D.3 Organ-culture dish. A 35-mm petri dish is cemented into the bottom of a 60-mm petri dish. The 35-mm dish is filled with about 2 ml of albumen–agar. Before use, cotton or lab tissue that has been moistened with Howard–Ringer's solution is placed in the area inside the larger dish but outside the 35-mm dish.

3. When Eggs Are Delivered

a. Hold them at 15–18°C until the appropriate time for incubation. The eggs should be rotated twice daily after their arrival.

b. Start incubation at the appropriate time before lab by transferring the desired number of eggs from 15 to 37°C.

c. For Part II.F, determine empirically how long eggs need to be incubated to reach the 6–7 somite stage. To do so, incubate 6–12 eggs. About 24 h after the beginning of incubation, dissect and stage one of the embryos. Continue to dissect one of the eggs every 3 h until an embryo having 6–9 somites is found (usually 8–15 additional hours of incubation). Dissect several additional embryos to confirm that the incubation time is correct. Start incubation of eggs for Part II.F at this time prior to the lab.

4. The Day before the Laboratory Period

a. Prepare and autoclave Howard–Ringer's (or chick Ringer's) solution according to Table D.4. Autoclave the solution and cool it at least to about 60°C, and then add 60 mg/liter of penicillin G (1625 I.U./mg) and 50 mg/liter streptomycin sulfate.

b. Prepare 70% ethanol for students to sterilize their instruments with.

c. Cut at least two "doughnuts" per student from No. 1 filter paper, about 25 mm in outer diameter, about 1.75 mm in inner diameter; wrap them in foil, and autoclave to sterilize.

d. Plug Pasteur pipettes with cotton. Prepare several packages, each containing about 25 plugged Pasteur pipettes wrapped in foil, and autoclave to sterilize.

e. Prepare a candling apparatus by cutting a hole about 1 in. in diameter in a small box and taping the box to a flashlight with its hole positioned over the flashlight's beam (see Fig. D.4).

TABLE D.4

Quantities for Preparation of Howard–Ringer's (or Chick Ringer's) Solution

Chemical	Formula weight (g/mol)	g/Liter	Molarity (mM)
NaCl	58.44	7.0	120.0
CaCl$_2$–2-H$_2$O	147.2	0.32	2.2
KCl	74.5	0.42	5.6
H$_2$O	—	(To 1 liter)	—

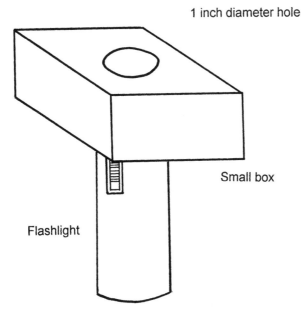

FIGURE D.4 Apparatus for candling chicken eggs. A small box with a 1-in.-diameter hole cut in it is taped to a flashlight. Place eggs to be candled over the hole and examine for the desired feature, such as an embryo or blood vessels.

5. The Day before Part II.F

a. Finish preparation of the organ-culture dishes by pipetting albumen–agar into the inner 35-mm dish of each. To prepare albumen–agar:

(1) Mix 4 g agar, 4 g glucose, and 100 ml Howard–Ringer's solution in a 250-ml flask, cap loosely, and autoclave 20 min using slow exhaust cycle. Cool to about 60°C.

(2) Collect 300 ml of albumen (from about five fresh eggs). Stir gently to disperse clumps. Warm to 45°C in a water bath.

(3) After agar solution has cooled, mix it with the albumen. Keep it in the water bath to prevent it from solidifying.

(4) Using sterile technique, pipette 2 ml of the mixture into the inner 35-mm cup of each organ-culture dish.

(5) After albumen–agar has solidified, store the plates in a moist chamber in the refrigerator until use.

6. The Day of Lab

   a. Cover the surface of hot plate and surrounding lab bench with foil to protect it from wax drips, and turn it on at lowest heat. Place a cube of wax (e.g., Gulfwax) in a Pyrex glass dish to melt on the hot plate.
   b. Set out the following supplies for each lab bench:

250-ml beakers of 70% alcohol for student instrument sterilization
Squirt bottles of 70% ethanol
Kimwipes
Rulers
Files
Pencils
Sterile Pasteur pipettes
Bottle of Howard–Ringer's
22-×-22-mm coverslips
Wide clear packing tape
Capillary tubes for preparing glass knives
Bunsen burner and striker
Petri dishes to hold eggs during windowing

   c. Set out the prepared slides.

C. Resources

1. Fertile Eggs

   Fertile eggs can be obtained from local or national suppliers. Consult the nearest agricultural extension station for local or regional suppliers.

Carolina Biological Supply Company
2700 York Road
Burlington, NC 27215
Telephone: (919) 584-0381

Truslow Farms, Inc.
Route 4, Box 118
Chestertown, MD 21620-2706
Telephone: (410) 778-3000

Ward's Natural Science Establishment
5100 West Henrietta Road
P.O. Box 92912
Rochester, NY 14692-9012
Telephone: (176) 359-2502

2. Incubators

Purchase of a humidity- and temperature-controlled rotating incubator is highly recommended. Information can be obtained from the nearest Agricultural Extension Station. Possible sources include:

Humidaire Incubator Co.
217 W. Wayne St.
New Madison, Ohio 45346-9791
Telephone: (937) 996-3001

Lyon Electric Co.
Box 81303
San Diego, CA 92138

Stromberg's Chicks and Pets Unlimited
Box 400
Pine River, MN 56474
Telephone: (218) 587-2222

3. Prepared Slides

Prepared slides can be purchased from:

Carolina Biological Supply Company
2700 York Road
Burlington, NC 27215
Telephone: (919) 584-0381

Ward's Natural Science Establishment
5100 West Henrietta Road
P.O. Box 92912
Rochester, NY 14692-9012
Telephone: (176) 359-2502

V. GUIDE TO PREPARATION, LABORATORY EXERCISE 4: *DROSOPHILA* GENE EXPRESSION

A. Equipment and Supplies List

1. Equipment for Preparation of Lab

Bottles and/or vials for culturing flies
Apparatus for anesthetizing flies (a large gauze-wrapped cotton wad dampened with ether)

Morgue for discarding adult flies

16–20°C incubator for growth of larvae for polytene chromosome squashes

37°C incubator for heat-shock treatment of flies

Fine watercolor paintbrushes

2. Chemicals and Materials for Preparation of the Lab

Standard: Acetic acid, ether

Special: *Drosophila* food, natural orcein stain, nail polish to seal coverslip edges, microscope slides and coverslips, Kimwipes, two pairs of fine forceps or an insect dissecting pin and a one pair of forceps, depression slides (if available), 1.5-ml plastic tubes for sample preparation, sample buffer for gel electrophoresis, SDS–PA gels, apparatus, buffers, and power supply

Live animals: 1- to 2-week-old cultures of *Drosophila*

B. Timeline

1. Several Months before the Date of the Lab

a. Obtain cultures of flies suitable for the lab. Although D. *melanogaster* work fine, the salivary glands of larger species such as D. *virilis*, D. *pseudoobscura*, and D. *mojavensis* are larger and give better results for beginners. Larvae grown at 16–20°C are usually larger and easier for beginners to dissect.

b. Expand the cultures into fresh bottles. To do so, sex the adult flies, and put 4–6 pairs in each bottle of fresh medium. Males are generally distinguished from females by their smaller size, more rounded and darkly pigmented rear ends, and dark dense sex combs on the backs of their first pair of legs.

To transfer and sex adult flies:

(1) Knock them to the bottom of the culture vessel by gently rapping the vial on the palm of your hand or a padded surface, such as a computer mouse pad.

(2) Quickly replace the culture-bottle stopper with a gauze-wrapped cotton ball dampened with several drops of ether (take care not to over-etherize the flies).

(3) Turn the culture vessel on its side to prevent the etherized flies from sticking to the medium in the bottom of the vial.

(4) Watch for the flies to stop moving. Remove the ether-containing stopper, and dump the flies out onto an index card or other smooth white surface.

(5) Examine the flies using the dissecting microscope to determine their sex, sorting them into piles of males and females using a small paint-

brush, closed fine forceps, or a dissecting tool. Work quickly so that flies are sorted before they recover from anesthesia.

(6) Put 4–6 pairs of adult flies into each bottle of fresh medium, and cover with a stopper.

(7) Discard unwanted adult flies into a can partially filled with mineral oil (morgue).

c. Maintain culture bottles at room temperature for several days, and then transfer the adults to fresh culture bottles. Retain the emptied bottle for the larvae, pupae, and fresh adults that will form over the next weeks. Depending on the class size, several cycles of expansion may be necessary.

2. One to Two Weeks before the Lab

a. Expand the cultures into enough fresh vessels for each student or pair of students in the class to have one.

b. Maintain these cultures at 16–20°C, with plenty of food, and somewhat more moist than cultures used for stock maintenance, expansion, and crosses.

c. Prepare the 2% aceto-orcein stain:

3 g of natural orcein
75 ml glacial acetic acid
30 ml lactic acid
45 ml distilled H_2O

Mix the above reagents. Heat gently without boiling. Filter twice through No. 1 filter paper. When cool, distribute into small bottles with tight-fitting stoppers or caps.

3. The Day before Lab

a. Prepare SDS–PA gels and buffer solutions for the class (see Appendix C).

b. Equilibrate an incubator at 37°C.

c. Prepare 45% acetic-acid fixative.

4. The Day of Lab

a. Remove the adult flies from vials for student use. Reserve the adult flies from at least one vial for student observation.

b. At each lab bench, set out:

Vials of eggs, larvae, and pupae reared at 16–19°C
Droppers and bulbs for fixative and stain
45% acetic-acid fixative
2% aceto-orcein stain

Depression slides
Microscope slides and coverslips
Kimwipes
Bottles of nail polish

c. Prepare a boiling-water bath to denature protein gel samples after students prepare them. Prepare to run the SDS–PA gels (see Appendix C).

C. Resources

1. Drosophila Cultures

Obtain cultures of *Drosophila* from colleagues who work with flies, or from the suppliers listed below.

Carolina Biological Supply Company
2700 York Road
Burlington, NC 27215
Telephone: (919) 584-0381

Ward's Natural Science Establishment
5100 West Henrietta Road
P.O. Box 92912
Rochester, NY 14692-9012
Telephone: (176) 359-2502

2. Instant Drosophila Medium

Instant *Drosophila* medium can be obtained from Carolina Biological Supply Company (see address above) and is simple and easy to prepare for short-term maintenance of cultures used in this lab.

3. Alternative Anesthesia

Flies can also be anesthetized with commercially available Flynap, obtained from Carolina Biological Supply Company, rather than ether as suggested above.

4. Natural Orcein

Natural orcein is available from:

Sigma Chemical Company
P.O. Box 14508
St. Louis, MO 63178
Telephone: (800) 325-3010

VI. GUIDE TO PREPARATION, LABORATORY EXERCISE 5: AMPHIBIAN METAMORPHOSIS

A. Equipment and Supplies List

1. Equipment for Preparation of Lab

Eight containers for maintaining tadpoles (preferably clear plastic containers that can be placed directly on the stage of a dissecting microscope for tadpole measurement)

Well water or tap water dechlorinated by at least 12 h of exposure to air

500-ml graduated cylinders for preparing solutions

2. Chemicals and Materials for Preparation of the Lab

Standard: none

Special: 8 dipnets, millimeter rulers, formaldehyde, thyroxine–5-H_2O, iodine, actinomycin D, 15-ml disposable plastic tubes or flasks for fixing dead tadpoles, Pasteur pipettes and bulbs, tadpole food

Live animals: tadpoles at the 1-mm hind-limb stage (40 or more)

3. Equipment and Supplies Needed by Each Student (See Materials Section of Laboratory Exercise 5)

B. Timeline

1. Before the Course Begins

a. Decide whether embryos from Laboratory Exercise 2 (Early Amphibian Development) will be reared and used for this lab or whether tadpoles at the hind-limb stage will be obtained from another source. Tadpoles can be collected from local ponds or streams, although this source is highly dependent on seasonal weather conditions. Tadpoles can also be obtained from commercial supply houses (see list below). Tadpoles of species other than *X. laevis* can be used (such as *Rana*), provided they are at the 1-mm hind-limb stage.

2. When Tadpoles Hatch and Start Feeding (or Arrive)

a. Develop a feeding and care schedule for students to follow over the next weeks in feeding and caring for the tadpoles and in collecting growth data.

b. Tadpoles can be fed frozen or canned spinach or collard greens or parboiled lettuce. Monitor the care and feeding of tadpoles, and avoid overfeeding (see Laboratory Exercise 5).

3. When Tadpoles Reach the 1-mm Hind-Limb Stage

a. Divide the tadpoles into eight groups, and place them in separate containers of shallow water.

b. Revise the feeding and care schedule, dividing the students into eight groups and assigning each student group care of one group of tadpoles.

c. Prepare stock solutions of thyroxine, iodine, actinomycin D, and 5% formaldehyde fixative as follows:

0.1 mM thyroxine stock solution: Dissolve 35.5 mg of thyroxine–5-H_2O in 400 ml of distilled water. Store stock solution at 4°C.

0.1 mM iodine stock solution: Dissolve 5.08 mg of iodine in 200 ml of distilled water. Store stock solution at 4°C.

0.25 mg/ml actinomycin D stock: Dissolve 10 mg of actinomycin D in 40 ml of distilled water. Store stock solution at 4°C.

5% formaldehyde fixative solution: In a fume hood, dilute 13.5 ml 37% formaldehyde to 100 ml with distilled water. Store tightly covered at room temperature, and avoid breathing of fumes or contact with skin or eyes.

C. Resources

1. Tadpoles

Tadpoles can be obtained commercially from:

Carolina Biological Supply Company
2700 York Road
Burlington, NC 27215
Telephone: (919) 584-0381

Ward's Natural Science Establishment
5100 West Henrietta Road
P.O. Box 92912
Rochester, NY 14692-9012
Telephone: (176) 359-2502

2. Chemicals

Thyroxine, iodine, actinomycin D, and concentrated formaldehyde can be obtained from:

Sigma Chemical Company
P.O. Box 14508
St. Louis, MO 63178
Telephone: (800) 325-3010

3. Other Materials

Plastic containers, dipnets, and tadpole food can be obtained locally.

VII. GUIDE TO PREPARATION, LABORATORY EXERCISE 6: CELL-CELL INTERACTIONS DURING SPONGE AGGREGATION

A. Equipment and Supplies List

1. Equipment for Preparation of Lab

Seawater tanks for holding the sponges until lab day
Two large beakers
4°C incubator
Two rotary shakers (one at room temperature and one at 4°C)
Ice buckets for student and preparator use
Tongs
Refrigerated centrifuge
Centrifuge tubes

2. Chemicals and Materials for Preparation of the Lab

Standard: NaCl, KCl, $MgCl_2$-6-H_2O, $MgSO_4$-7-H_2O, $CaCl_2$-2-H_2O, $NaHCO_3$, EGTA

Special: glucose, galactose, glucosamine, glucuronic acid, galacturonic acid, knife or razor blade for chopping sponges, gauze or cheesecloth, waterproof gloves (optional)

Live animals: sponges of two different genera (and colors)

3. Equipment and Supplies Needed by Each Student (See Materials Section of Laboratory Exercise 6)

CAUTION: Some individuals are allergic to sponges. Take care either to use tongs or, if you use your fingers to disperse the cells, to wear gloves.

B. Timeline

1. Several Months before the Lab

a. Set up the seawater tanks to hold the sponges as for the Laboratory Exercise 1 (see Appendix D.II).

b. Arrange for shipment of two different genera of sponges of different colors to be delivered as close to the lab date as possible.

2. When the Sponges Arrive

a. Clear the seawater tank of other predatory occupants, such as sea urchins or crabs, and place the sponges in the tank.

3. The Day before Lab

a. Prepare artificial seawater and CaMgFSW, as for Laboratory Exercise 1. Artificial Seawater (SW): Use fresh synthetic sea salts, such as Instant Ocean, or prepare according to Table D.1. Adjust pH to 8.0–8.2 and specific gravity to 1.020–1.023 at 75°C. Add 10^5 U/liter penicillin and 100 mg/liter streptomycin to increase viability of the embryos.

b. Prepare the sugar solutions according to Table D.5.

c. Prepare a stock solution of 0.1 M $CaCl_2$: Dissolve 1.47 g $CaCl_2$–2-H_2O in 100 ml distilled H_2O.

d. In the late afternoon or evening before lab, begin preparation of aggregation factor (AF):

(1) Weigh sponge tissue (e.g., 2.5 g of the red *Microciona* sponge and 10 g of the *Cliona* sponge).
(2) Rinse three times in 10 ml of cold, fresh CaMgFSW.
(3) Place the pieces of one species of sponge in a square of gauze three layers thick and gather up the edges of the gauze to make a bag.
(4) Dip the bag into a small beaker containing 25 ml of cold CaMgFSW and squeeze the bag to disperse the cells.
(5) Repeat the previous two steps to disperse cells of the other type of sponge.
(6) Secure the beakers on a rotary shaker and swirl at 60–80 rpm at 4°C overnight.

TABLE D.5

Preparation of Sugar Solutions

Sugar	Formula weight (g/mol)	g/300 ml	Final molarity (M)
Glucose (dextrose)	180.16	27.0	0.5
Galactose	180.16	27.0	0.5
Glucuronic acid, N^+ salt[a]	216.1	32.4	0.5
Galacturonic acid–H_2O[a]	212.2	31.8	0.5
Glucosamine	215.6	32.3	0.5

Note. Dissolve in SW.
[a]Neutralize with 0.1 N NaOH.

4. The Day of Lab

a. Set out for each student:

Two 12-well multiculture dishes
Hemacytometers
Pipettes and bulbs
Bottles of each type of sugar solution and seawater
Ice buckets for keeping dispersed cells and AF cold until use
Dissecting and compound microscopes

b. Finish preparation of AF

(1) Transfer the sponge solutions from the beakers into 50-ml round-bottomed centrifuge tubes and spin at low speed (1000 rpm in a Sorvall SS-34 rotor for 2 min).

(2) Decant the supernatant into a clean centrifuge tube and spin at high speed (17,000 rpm in a Sorvall SS-34 rotor for 20 min).

(3) Decant the supernatant into a clean tube and measure its volume. Add 0.1 M $CaCl_2$ to a final concentration of 3.0 mM (add 0.3 ml of 0.1 M $CaCl_2$ for every 10 ml of AF solution).

(4) Store the AF on ice until use.

c. Prepare to disperse sponges for the lab. To disperse the sponge cells, wash the sponges several times with fresh SW. Using a knife, scalpel, or razor blade and glass cutting board, chop the sponge tissue into cubes 0.5–1 cm on a side. Place pieces of both sponge species in a large square of gauze three or more layers thick, gather the edges to make a bundle, dip the bundle into a beaker of seawater, and squeeze the bag to disperse the cells. Maintain the dispersed cells on ice during the lab session.

C. Resources

Sponges can be obtained from local pet stores that specialize in seawater aquaria and animals. They can also be obtained from commercial sources listed below. Call the supplier to determine availability of sponges of different colors when needed. Usually, one sponge about the size of a tennis ball will supply enough dispersed cells for 25 students.

Marinus, Inc.
1500 Pier C Street
Long Beach, CA 90813
Telephone: (310) 435-6522

Pacific Bio-Marine Laboratories, Inc.
P.O. Box 536
Venice, CA 90291
Telephone: (213) 822-5757

Gulf Marine Specimens
P.O. Box 237
Panacea, FL 32346
Telephone: (850) 984-5297

Animal Supply House
Marine Biological Laboratory
Woods Hole, MA 02543
Telephone: (617) 548-3705, ext. 325

Sugars can be obtained from:

Sigma Chemical Co.
P.O. Box 14508
St. Louis, MO 63178
Telephone: (800) 325-3010

VIII. GUIDE TO PREPARATION, APPENDIX B: MICROSCOPE CARE AND USE

A. Equipment and Supplies List

1. Equipment for Preparation of Lab

Hemacytometers
Plastic ruler fragments (2–3 in. long)
Dissecting and compound microscopes

2. Chemicals and Materials for Preparation of the Lab

Standard: microscope slides and coverslips, Kimwipes
Special: toothpicks, NaCl, $CaCl_2$-2-H_2O, Tris, HCl, bean seeds, mustard or radish seeds, celery seeds, onion root tips, immobile *Drosophila* adults of both sexes, embryo whole mounts (18- to 96-h chick embryos or frog gastrula sections), Vaseline, Kimwipes, lens paper
Live materials: human cheek cells

B. Timeline

1. Before the Course Begins

      a. Gather the supplies needed for the lab. Many of these supplies (such as hemacytometers, microscopes, ruler fragments) are used in several other labs, so they should be kept on hand for the remainder of the semester.

      b. Many of the reagents for the lab are nonperishable and can be gathered at any time prior to the lab. Saline solution for cheek cells can be prepared at any time and refrigerated until use.

2. The Day before the Lab Period

      a. Prepare saline solution according to Table D.6.

      b. Set out measuring tools and objects to be measured at each lab bench.

Bean, celery, radish, and mustard seeds
Slide boxes
Toothpicks
Vaseline
Microscope slides and coverslips
Cheek saline and droppers
Kimwipes
Ruler fragments
Hemacytometers

TABLE D.6

Preparation of Saline Solution

Chemical	Formula weight (g/mol)	g/0.25 l	Final molarity (mM)
NaCl	58.44	2.2	150
$CaCl_2-2-H_2O$	147.2	0.37	5
Tris base	121.2	0.30	10

Note. Dissolve chemicals in 200 ml distilled H_2O; adjust pH to 7.5 with HCl and volume to 250 ml.

IX. GUIDE TO PREPARATION, APPENDIX C: GEL ELECTROPHORESIS

Two types of gel electrophoresis are demonstrated in this lab, SDS–polyacrylamide gel electrophoresis and agarose gel electrophoresis. Instructions for preparation of gels for the two types are described separately below.

A. Equipment and Supplies List: SDS–Polyacrylamide Gel Electrophoresis

1. Equipment for Preparation of Lab: SDS–Polyacrylamide Gel Electrophoresis

CAUTIONS: Unpolymerized acrylamide is a neurotoxin. Wear gloves when handling it. Do not attach or detach the leads from the gel apparatus with the power supply on. Do not dip your fingers in the buffer tanks while gel is running.

Vertical gel apparatus for running SDS–PA gels (see Fig. C.1)
Glass plates, spacers, and comb for casting vertical gels (see Fig. C.1)
Boiling-water bath
Heat-resistant test tube rack
1.5-ml plastic tubes with attached lids for sample preparation
Power supply
Micropipettors
Disposable tips for micropipettors

2. Chemicals and Materials for Preparation of the Lab: SDS–Polyacrylamide Gel Electrophoresis

Coomassie blue stain, acetic acid, methanol, ammonium persulfate, TEMED, DTT, acrylamide, bis-acrylamide, Tris base, HCl, glycine, SDS, bromophenol blue or phenol red dye, molecular-weight markers

B. Timeline: SDS–Polyacrylamide Gel Electrophoresis

1. At Any Time before the Lab

a. Prepare the following chemicals and stock solutions.

30-0.8% acrylamide: Dissolve 30 g acrylamide and 0.8 g bis-acrylamide in 80 ml distilled H_2O and bring to 100 ml volume. Store in amber bottle at 4°C.
1 M Tris-Cl, pH 6.8: Dissolve 12.11 g Tris base in 60 ml distilled H_2O, add drops of concentrated HCl until the pH is 6.8, and adjust the final volume to 100 ml with distilled H_2O. Store at 4°C.

3 M Tris-Cl, pH 8.8: Dissolve 36.33 g Tris base in 60 ml distilled H_2O, add drops of concentrated HCl until the pH is 8.8, and adjust the final volume to 100 ml with distilled H_2O. Store at 4°C.

10% SDS: Dissolve 20 g sodium dodecyl sulfate (SDS) in 80 ml distilled H_2O, and adjust the volume to 100 ml. Store at room temperature.

100% TEMED: Purchase as a liquid stock. Store at 4°C.

1g /10mLs

0.1g/mL

10% ammonium persulfate: Dissolve 50 mg in 0.5 ml of distilled H_2O just before use.

1× Tris-glycine running buffer: Dissolve 3.03 g Tris, 14.4 g glycine, and 1 g SDS in 800 ml distilled H_2O, and adjust the pH to 8.3 with HCl and the volume to 1 liter. Prepare just before use.

Staining solution: Dissolve 1 g Coomassie brilliant blue R dye in 1 liter of solution 25% methanol, 10% glacial acetic acid. Filter through No. 1 filter paper, and store at room temperature.

Destaining solution: Mix 250 ml methanol, 100 ml glacial acetic acid, and 650 ml distilled H_2O. Store at room temperature.

2 ×; SDS sample buffer: Dissolve 1.94 g Tris base in 50 ml distilled H_2O and adjust the pH to 6.8 with HCl. Add 4 g SDS, 3.08 g dithiothreitol, 20 ml glycerol, and enough bromophenol blue or phenol red dye to give a deep blue or red color; adjust the volume to 100 ml. Store frozen at −20°C in 10-ml aliquots.

Molecular-weight markers: Purchase molecular-weight markers for the gels. A convenient range of protein molecular weights includes:

Protein	Molecular weight (Da)
phosphorylase B	110,000
bovine serum albumin	84,000
ovalbumin	47,000
carbonic anhydrase	33,000
trypsin inhibitor	24,000
lysozyme	16,000

2. The Day of Lab

a. Early in the morning, prepare gels for use in lab later in the day.

(1) Slab gels are ideal for use in the labs in this manual. Follow the manufacturer's instructions for assembling the gel plates of equipment already on hand. Homemade equipment of any desired size can be assembled as shown in Figs. C.1 and C.2.

(2) Clean all grease and dirt from the plates. Assemble plates using spacers to which a bead of vacuum grease has been applied on both flat sides as shown in Fig. C.2A. Hold the spacers and plates together with butterfly clamps along the sides and bottom of the plates (see Fig. C.2A).

10%
Resolving gel

(3) Pour a 10% resolving gel (gives generally good resolution of most complex protein samples). In a 250-ml sidearm flask, mix:

16.6 ml 30-0.8% acrylamide stock
6.25 ml 3 M Tris-Cl, pH 8.9, stock
5 ml 10% SDS stock
27 ml distilled H_2O, to make 50 ml.
Swirl to mix, and degas the solution by attaching the stoppered flask to a vacuum line for 2 min. Then add
30 μl TEMED stock
300 μl 10% ammonium persulfate stock
Swirl gently, and pour the mixture between the assembled plates to within 1.5 cm of the notch in the gel plate. Slowly and gently overlay the acrylamide mixture with a 2- to 3-mm layer of water. Wait 15–30 min for a sharp interface to appear between the water layer and the acrylamide.

(4) Drain as much of the water overlay as possible from the top of the gel. Prepare a 4.5% stacking gel by mixing:

4.5 ml 30-0.8% acrylamide stock
3.75 ml Tris-Cl, pH 6.8
300 μl 10% SDS
Add 21.45 ml distilled H_2O and swirl to mix. Degas under vacuum for 2 min, and add
30 μl TEMED
50 μl 10% ammonium persulfate
Pour this mixture on top of the resolving gel, slide comb into position, gently overlay with water, and wait at least 15–30 min for the stacking gel solution to polymerize.

(5) After the stacking gel has polymerized and samples have been prepared (see below), remove the bottom spacer from the gel plates. Apply a bead of vacuum grease along the outside of the notched gel plate to seal the gel plates against the apparatus, and clamp the gel plates to the apparatus with butterfly clamps (see Fig. C.2).

(6) Fill the tanks of the apparatus with Tris-glycine running buffer so that the buffer level is above the notch in the back gel plate.

b. Sample preparation and gel loading

(1) Collect cell or embryonic samples. For a lab introducing gel electrophoresis, students can scrape their inner cheeks with toothpicks to obtain a sample of about 10 μl and deposit it in a 1.5-ml plastic centrifuge tube with attached lid. Add 10–20 μl of 2× sample buffer using a microliter pipettor, and pipette up and down several times to mix. Close the tube and place it in the rack. Boil the samples for 5 min, remove them from water bath, and allow them to cool slightly.

(2) Gently remove the comb from the gel–plate sandwich.

(3) Load one sample into each well of the stacking gel (see Fig. C.2B). Reserve one lane for loading the molecular-weight markers. Note in your lab notebook which sample is loaded into each well.

(4) Plug the apparatus into the power supply and run the gel at 30–40 mA, constant current, until the sample-buffer dye has run to the bottom of the gel (see Fig. C.2C).

(5) Turn off the power supply and unplug the apparatus from it.

c. Gel staining and destaining

(1) Remove the running buffer from the upper and lower reservoirs of the gel apparatus. Remove the gel plates from the apparatus by releasing the butterfly clamps.

(2) Remove the spacers from the sides and bottom of the "sandwich." One of the spacers can be partially removed and used to pry the plates apart. The gel will usually stick to one of the plates.

(3) Transfer the gel from that plate to a staining tray large enough to accommodate the gel without folding it and containing enough staining solution to cover the gel; agitate it gently on a rotary shaker. When the gel is sufficiently stained, pour off the staining solution and replace it with destaining solution. Continue gentle agitation until the background of the gel is clear and protein bands are visible. Times for staining and destaining depend on the thickness and the concentration of the gel.

(4) After destaining, the gel can be dried onto filter paper or between sheets of porous plastic film for preservation.

A. Equipment and Supplies List: Agarose Gel Electrophoresis

1. Equipment for Preparation of Lab: Agarose Gel Electrophoresis

CAUTIONS: Ethidium bromide is a mutagen. Wear gloves when handling it. Place any contaminated wet or dry waste in the container designated for hazardous waste. Do not look directly at the UV lamp without safety goggles or a face shield. Limit exposure of your skin to UV light. Do not attach or detach the leads from the gel apparatus with the power supply on. Do not dip your fingers in the buffer tanks while gel is running.

Horizontal gel apparatus for agarose gels (see Fig. C.3)
Agarose gel casting tray and comb (see Fig. C.3)
Tape
Hand-held UV source
UV-protective eyewear
Power supply

Micropipettor
Disposable tips for micropipettor
1.5-ml plastic sample tubes

2. Chemicals and Materials for Preparation of the Lab: Agarose Gel Electrophoresis

Tris base, Na acetate–2-H_2O, 2Na–EDTA, ethidium bromide, agarose, plasmid DNA, three different restriction endonucleases, DNA molecular-weight markers

B. Timeline: Agarose Gel Electrophoresis

1. At Any Time before the Lab

(a) Prepare the following chemicals and stock solutions.

50× TAE buffer: Dissolve 242.2 g Tris base, 20.5 g Na acetate–3-H_2O, and 37.2 g EDTA–2-H_2O in 800 ml distilled H_2O, adjust pH to 8.0 with acetic acid and volume to 1 liter with distilled H_2O. Store at 4°C.

10× stop buffer: Dissolve 15 g Ficoll (MW 400,000), 7.4 g EDTA, and 100 mg bromophenol blue and adjust the volume to 100 ml with distilled H_2O. Store at 4°C.

1% agarose in 1× TAE buffer: In a 250-ml flask, dissolve 1.5 g of agarose in 100 ml of 1× TAE buffer with gentle heat by stirring on a hot plate or with frequent swirling in a microwave oven. Take care not to let the mixture boil out of the flask. If gel will be poured immediately, cool the molten agar to 65°C before addition of ethidium bromide stock and pouring onto gel tray. Agarose can be prepared in advance and remelted before use.

10 mg/ml ethidium bromide stock: Dissolve 100 mg of ethidium bromide in 10 ml distilled H_2O. Store in amber bottle at room temperature.

DNA molecular-weight markers: Purchase predigested DNA from a commercial supplier and suspend it in 1× stop buffer. Bacteriophage lambda DNA digested with the restriction endonuclease *Hind*III produces DNA fragments of sizes suitable for separation on a 1% gel: 23,100, 9400, 6600, 4400,* 2300, 2000, 600,* 100.* Asterisks mark fragments whose bands are difficult to see.

(b) Prepare DNA samples for gel electrophoresis.

Digest plasmid DNA with restriction endonucleases according to the supplier's instructions. In general, 0.25 μg of plasmid DNA cut into 2–3 bands can be visualized easily in a single lane of a gel. Check that the DNA was digested completely by running part of the sample on a gel before lab. A suggestion is to digest, according to the following scheme, enough of the plasmid pGEM3Zf+ that each student can load several lanes if desired: (i) uncut, no enzyme added; (ii) cut 1× with *Bam*HI endonuclease to produce a linear fragment of 3000 base pairs (bp); (iii) cut with *Bam*HI and *Bgl*I to generate three fragments of 1574, 1458, and 194 bp; (iv) cut 2× with *Bgl*I to generate two fragments of 1574 and 1458. After digestion, dilute the sample to 0.016 mg/ml with TAE buffer, add 1 (new) vol of 10-stop buffer, mix, and store at 4°C.

2. The Day of Lab

a. Melt the agarose if it was prepared ahead, or prepare fresh 1% agarose before lab. Prepare 1× TAE running buffer by dilution of the 50× stock.

b. Seal the edges of the gel-casting tray with tape. Adjust the comb height so that the bottom of the wells is about 1 mm thick (see Fig. C.4B).

c. Add ethidium bromide to gel mix that is cooled to 65°C, swirl gently, and pour the gel mix into the casting tray to form a gel that is about 3–4 mm thick (see Fig. C.4A). The agarose gel will harden as it cools.

d. Carefully remove the comb from the solidified gel, remove the tape from around the ends of the gel, place the gel in the horizontal gel apparatus, and fill the tank with sufficient 1× TAE buffer just to cover the gel.

e. Load the samples, noting the order of loading.

f. Connect the electrode wires between the gel apparatus and the power supply. Run the gel at 75–100 V until the sample-buffer dye has traveled 2/3 the length of the gel.

g. Turn off the power supply, disconnect the electrodes, and place the gel still in its casting tray on a dark surface. Don the UV-protective eyewear and examine the gel using the hand-held UV light. The gel can be photographed in a darkroom for preservation.

C. Resources

1. Electrophoretic Supplies

SDS and agarose gel electrophoresis equipment and reagents can be obtained from many different general suppliers including Carolina Biological Supply; Fisher Scientific, Telephone: (800) 766-7000; Thomas Scientific, Telephone: (800) 345-2100; Baxter/Scientific Products, Telephone: (800) 342-0191; and Sigma Chemical Company. Specialty suppliers such as Bio-Rad [Telephone: (800) 424-6723] and Hofer sell equipment and reagents that are high quality and durable. Some companies supply gel reagent kits that require minimal preparation and are economical if agarose and SDS–PA gels are run infrequently. Some companies sell precast gels and samples for gel electrophoresis demonstrations.

2. Tape

Tape for sealing SDS and agarose gel plates can be obtained from 3M Company through VWR/Scientific Products Supply.

3. Restriction Endonucleases

Restriction endonucleases can be obtained from suppliers of molecular biology reagents:

GIBCO/BRL
Life Technologies, Inc.
P.O. Box 9418
Gaithersburg, MD 20898
Telephone: (800) 828-6686

New England Biolabs
32 Tozer Road
Beverly, MA 01915-5599
Telephone: (800) 632-5227

Promega Corporation
2800 Woods Hollow Road
Madison, WI 53711-5399
Telephone: (800) 356-9526

Stratagene
11011 North Torrey Pines Rd.
La Jolla, CA 92037
Telephone: (800) 424-5444

INDEX

Note: "t" closed up to page number indicates that item is in a table; "f" indicates figures.

A

Aggregation, cell-cell
  laboratory exercise, 47–52
  preparation for lab exercise, 101–104
Ambystoma mexicanum, see Axolotl
Amphibians
  laboratory exercise, 13–20, 41–45
  preparation for lab exercise, 84–89
  sources, 88–89
  timetable for development, 15t
Arbacia punctulata, see Sea urchins
Axolotl, *Ambystoma mexicanum*
  early development, 13, 19–20
  sources, 89

B

Blastomere separation, 16–17, 20

C

Cardia bifida, 27–31
Cell death, programmed, 26, 31
Chick, *see* Chicken
Chicken, *Gallus domesticus*
  laboratory exercise, 21–31
  preparation for laboratory exercise, 89–95
  Ringer's solution for, preparation, 92t
  sources, 94
Chromosomes, polytene, 33–40
Cleavage plane reorientation. 17–18, 20

D

Cliona, see Aggregation, cell-cell
Cortical rotation, 16, 20

D

Development, amphibian, *see* Amphibian development
Drosophila, see Gene expression

E

Electrophoresis, gel
  laboratory exercises, 61–71
  preparation for exercises, 106–112
  study of gene expression, 37, 40
Embryo inversion, 16

F

Fertilization
  cross-species, 7–8, 10
  sea urchin
    laboratory exercise, 1–11
    preparation for exercise, 76–84
Fertilization membrane, 3f, 9
Field diameter, determination, 56–58
Frogs, *see* Amphibians

G

Gallus domesticus, see chicken
Gene expression
  Drosophila
    laboratory exercise, 33–40

Gene expression *(continued)*
    preparation for laboratory exercise, 95–98
    sea urchin, 5–9, 10–11
Grafting, tissue, 25–26, 31

H

Howard-Ringer's solution, preparation, 92t

J

Jelly coats, 2, 3f, 6, 9

M

Metamorphosis, amphibian
    laboratory exercise, 41–45
    preparation for laboratory exercise, 99–101
Microciona, see Aggregation, cell-cell
Microscopes
    care and use, 55–60
    preparation of laboratory exercise on care
        and use, 104–105

N

Notebook, laboratory, 53–54

P

Parthenogenetic activation, 3, 8, 10, 14, 18, 20
Plasmid mapping, 70–71
Polyspermy, 7, 10
Polytene chromosomes, *see* Chromosomes

R

Readings, supplementary
    amphibian metamorphosis, 45
    cell-cell interactions in sponges, 52
    chick embryonic development, 31
    early amphibian development, 20
    gene expression in *Drosophila*, 40
    sea urchin fertilization, 11
Regeneration, 25, 31
Reports, laboratory, 54

S

Saline solution, preparation, 105t
Sea urchins
    laboratory exercise on fertilization, 1–11

sources, 83–84
    timetable for embryonic development, 2t
Sea water, artificial, preparation, 78t
Size, microscopic, estimation, 56–59
Sources of supply
    anesthetics for *Drosophila*, 98
    animals, live, 83–84, 88–89, 98, 100,
        103–104
    chemicals, 84, 88, 100
    eggs, fertile chicken, 94
    electrophoresis supplies, 111
    endonucleases, 112
    incubators, 95
    medium for *Drosophila* cultures, 98
    nylon screen, 84
    orcein, natural, 98
    seawater–tank supplies, 84
    slides, prepared, 89, 95
    sugars, 104
    tape for preparation of electrophoresis gels,
        112
Sponges, *see also* Aggregation, cell–cell
    sources, 103–104
Steinberg's solution, preparation, 86t
Strongylocentrotus purpuratus, see Sea urchins
Sugar solutions, preparation, 102t

T

Threshold effects, 41
Thyroid hormones, 41

U

Urchins, *see* Sea urchins

V

Volume, microscopic, determination, 59–60

W

Windowing of chicken eggs, 23–25, 30

X

Xenopus laevis, see also Amphibians
    early development, 13–20
    metamorphosis, 41–45
    timetable for development, 15t